STUDENT WORKBOOK
MATH MADE VISIBLE

ELEMENTARY ALGEBRA FOR COLLEGE STUDENTS
NINTH EDITION

Allen Angel
Monroe Community College

Dennis Runde
State College of Florida

PEARSON

Boston   Columbus   Indianapolis   New York   San Francisco   Upper Saddle River
Amsterdam   Cape Town   Dubai   London   Madrid   Milan   Munich   Paris   Montreal   Toronto
Delhi   Mexico City   São Paulo   Sydney   Hong Kong   Seoul   Singapore   Taipei   Tokyo

⚠ This work is protected by United States copyright laws and is provided solely for the use of instructors in teaching their courses and assessing student learning. Dissemination or sale of any part of this work (including on the World Wide Web) will destroy the integrity of the work and is not permitted. The work and materials from it should never be made available to students except by instructors using the accompanying text in their classes. All recipients of this work are expected to abide by these restrictions and to honor the intended pedagogical purposes and the needs of other instructors who rely on these materials.

The author and publisher of this book have used their best efforts in preparing this book. These efforts include the development, research, and testing of the theories and programs to determine their effectiveness. The author and publisher make no warranty of any kind, expressed or implied, with regard to these programs or the documentation contained in this book. The author and publisher shall not be liable in any event for incidental or consequential damages in connection with, or arising out of, the furnishing, performance, or use of these programs.

Reproduced by Pearson from electronic files supplied by the author.

Copyright © 2015, 2011, 2008 Pearson Education, Inc.
Publishing as Pearson, 75 Arlington Street, Boston, MA 02116.

All rights reserved. No part of this publication may be reproduced, stored in a retrieval system, or transmitted, in any form or by any means, electronic, mechanical, photocopying, recording, or otherwise, without the prior written permission of the publisher. Printed in the United States of America.

ISBN-13: 978-0-321-92022-5
ISBN-10: 0-321-92022-8

1 2 3 4 5 6 EBM 17 16 15 14

www.pearsonhighered.com

PEARSON

Student Workbook For
ANGEL/RUNDE *Elementary Algebra for College Students*, 9e
Table of Contents

Chapter 1   1
    Section 1.2   1
    Section 1.3   6
    Section 1.4   10
    Section 1.5   12
    Section 1.6   14
    Section 1.7   17
    Section 1.8   19
    Section 1.9   22
    Section 1.10   26
    Vocabulary   29
    Practice Test A   31
    Practice Test B   33

Chapter 2   35
    Section 2.1   35
    Section 2.2   39
    Section 2.3   42
    Section 2.4   44
    Section 2.5   46
    Section 2.6   49
    Section 2.7   52
    Section 2.8   56
    Vocabulary   59
    Practice Test A   61
    Practice Test B   63

Chapter 3   65
    Section 3.1   65
    Section 3.2   69
    Section 3.3   72
    Section 3.4   75
    Vocabulary   79
    Practice Test A   81
    Practice Test B   83

Chapter 4   85
    Section 4.1   85
    Section 4.2   89
    Section 4.3   93
    Section 4.4   96
    Section 4.5   100
    Section 4.6   104
    Vocabulary   107
    Practice Test A   109
    Practice Test B   111

Chapter 5   115
    Section 5.1   115
    Section 5.2   117
    Section 5.3   119
    Section 5.4   121
    Section 5.5   123
    Section 5.6   125
    Section 5.7   127
    Vocabulary   129
    Practice Test A   131
    Practice Test B   133

Chapter 6   135
    Section 6.1   135
    Section 6.2   137
    Section 6.3   139
    Section 6.4   141
    Section 6.5   143
    Section 6.6   145
    Section 6.7   147
    Section 6.8   150
    Vocabulary   153
    Practice Test A   155
    Practice Test B   157

Chapter 7	**159**	**Answers**	**263**
Section 7.1	159	Chapter 1	263
Section 7.2	163	Chapter 2	265
Section 7.3	168	Chapter 3	267
Section 7.4	172	Chapter 4	268
Section 7.5	177	Chapter 5	269
Section 7.6	180	Chapter 6	270
Vocabulary	185	Chapter 7	272
Practice Test A	187	Chapter 8	275
Practice Test B	190	Chapter 9	277
		Chapter 10	279

Chapter 8 — **193**
- Section 8.1 — 193
- Section 8.2 — 197
- Section 8.3 — 199
- Section 8.4 — 201
- Section 8.5 — 204
- Vocabulary — 207
- Practice Test A — 209
- Practice Test B — 213

Chapter 9 — **217**
- Section 9.1 — 217
- Section 9.2 — 220
- Section 9.3 — 223
- Section 9.4 — 225
- Section 9.5 — 227
- Section 9.6 — 229
- Section 9.7 — 231
- Vocabulary — 233
- Practice Test A — 235
- Practice Test B — 237

Chapter 10 — **241**
- Section 10.1 — 241
- Section 10.2 — 243
- Section 10.3 — 245
- Section 10.4 — 248
- Section 10.5 — 251
- Vocabulary — 253
- Practice Test A — 255
- Practice Test B — 259

Name:
Instructor:
Date:
Section:

1.2 Problem Solving

Objectives
1. Learn the five-step problem-solving procedure.
2. Solve problems involving bar, line, and circle graphs.
3. Solve problems involving statistics.

Key Vocabulary
five-step problem-solving procedure, (algebraic) expression, operations, bar graph, approximately equal to ($\approx$), line graph, circle graph, measures of central tendency, mean, median, ranked data

1 Learn the five-step problem solving procedure.

Use and display the five-step problem-solving procedure to solve the following problems.

Example 1 A personal chef is preparing dinner for her Chicago client who is on a 1200 calories per-day restricted diet. If the client has already consumed 745 calories, mostly consisting of foods high in protein, how many calories can this client have for dinner?

2 Solve problems involving bar, line, and circle graphs.

Example 2 Some studies have shown that smoking rates vary depending on race, class or income. **Figure 1.1** compares the smoking rates of white and nonwhite mothers with varying educational backgrounds who smoked during pregnancy. Use the bar graph in **Figure 1.1** to answer the following questions.
  a) True or false: There are approximately twice as many mothers with less than 12 years of education who smoke during pregnancy than those with 12 years of education.
  b) True or false: No matter what the educational background, there are more white pregnant mothers who smoke than nonwhite pregnant mothers who smoke.
  c) In which educational background category is there the greatest percentage difference between white and nonwhite pregnant mothers who smoke? Approximate this percentage difference.

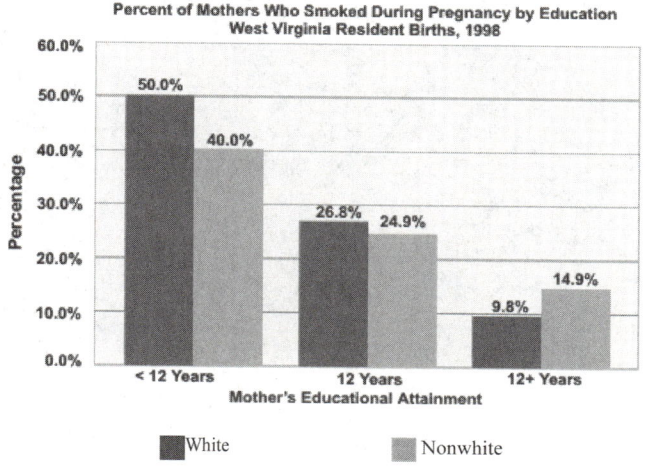

FIGURE 1.1 [Source: West Virginia Health Statistics Center]

Answers: 1. Step 1: The task is to find the number of calories the client has left to consume for dinner. The following information is given; yes (Y) if pertinent, no (N) if it is not pertinent: the chef is preparing dinner (N); client is in Chicago (N); client is on a 1200 calories per-day restricted diet (Y); client has already consumed 745 calories (Y); consumed foods high in protein (N). Step 3: calories allowed = 1200, calories consumed = 745, calories left = 1200 − 745 = 455 calories. Step 4: 455 + 745 = 1200; this is reasonable. Step 5: The client can consume 455 calories for dinner. 2a) true b) false c) The < 12 years education category has the greatest difference; the difference is 10%.

Copyright © 2015 by Pearson Education, Inc.   Angel/Runde, *Elementary Algebra for College Students*, 9e

Examples 1.2

Example 3 The breathtaking Australian Hotham Alpine Resort is located in the heart of the Victorian Alps. The line graph in **Figure 1.2** below summarizes the monthly snow accumulation in 2002, 2004, and 2006. Use the given graph to answer the following questions.
- a) During which year was there the most amount of snow accumulation?
- b) During which year was there the least amount of snow accumulation?
- c) Approximately how much snow accumulated between July and August in 2002?
- d) Approximately how much snow accumulated between July and August in 2006?

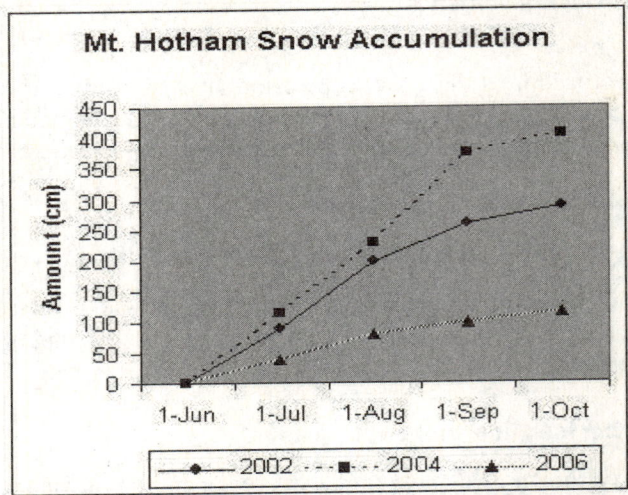

FIGURE 1.2   [Source: www.mthotham.com.au]

3 Solve problems involving statistics.

Example 4 Susan Renicke decided to take advantage of her utility company's budget billing. U-Lite-It Electric Company will average (calculate the mean of) the previous six bill amounts to determine the new payment with budget billing. Susan's last six bills were: $67.22, $43.04, $55.67, $89.22, $94.01, and $90.75. For Susan's bills, determine **a)** the average, or mean; **b)** the median; **c)** Susan's new payment.

Answers: 3a) 2004   b) 2006   c) approximately 100 cm   d) approximately 40 cm   4a) $73.32   b) $78.22   c) $73.32

Name:
Instructor:

Date:
Section:

Practice Set 1.2

1. The steps to Polya's five-step problem-solving procedure are listed below in random order. Arrange these steps with their corresponding number in the order in which they should be performed.

1) Carry out all necessary calculations.
2) Translate the problem to mathematical language.
3) Understand the problem.
4) Make sure you have answered the question.
5) Check the answer obtained in calculations.

1. _____

Use Polya's five-step problem-solving procedure to solve the following problems.

2. Corey Wiley has two coupons for his local Closets, Kitchens, and Things store. One coupon is for $5 off his total purchase and the other is for 20% off a single item. If he can only use one coupon to purchase a $16 mirror and a $5 set of pillow cases, answer the following questions.
a) Which coupon will save him the most money?
b) How much will he save?

2. a) _____

b) _____

3. The demand for genetically-engineered food crops grew between 1999 and 2001. In 1999 there were 6 million acres of crops for canola oil production and 48 million acres of soybeans. In 2001, there were approximately 4 million acres of crops for canola oil production and 82 million acres of soybeans.
a) Which crop had an increase in production and by how much?
b) Which crop had a decrease in production and by how much?

3. a) _____

b) _____

4. Sal Conway's test grades are 84, 65, 91, 70, and 75. For Sal's grades, determine the a) mean and b) median.

4. a) _____

b) _____

5. Faith Soellner's textbooks for her 1st semester of college cost $125.40, $67.35, $102.90 and $54.30. For the cost of these books, determine the a) mean and b) median.

5. a) _____

b) _____

Practice Set 1.2

Problem Solving

6. Michael Penn is pre-approved for a $12,000 car loan. His local car dealership is offering a 5% discount on its entire stock of cars.
a) If Michael finds a car for $12,600, can he afford it with the 5% discount?
b) What is the sale price?

6. a)_____

b)_____

7. Refer to #6 above. If the sales tax rate is 7%, how much sales tax will Michael pay?

7. _____

8. The AADT, or Average Annual Daily Traffic, is a measure of average daily traffic at specific locations. The AADT near Highway 47 and Highway N in 2004 was 2400 cars per day. The AADT at the same location in 1970 was 100 less than half the 2004 AADT. What was the AADT at this location in 1970?

8. _____

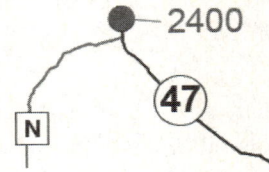

[Source: Missouri Department of Transportation]

9. The circle graph below shows the number of rainy, cloudy, and sunny days over a 2-week period. Use the graph to answer the following questions.
a) What fraction of the time was it sunny?
b) What percent of the time was it sunny?
c) What percent of the time was it rainy?

9. a) _____

b) _____

c) _____

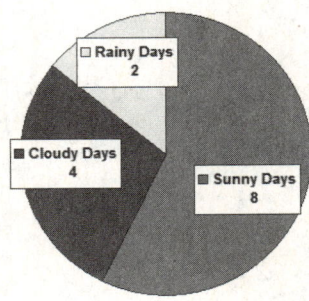

[Reprinted with permission of Stevens Institute of Technology; © 2007 the Trustees of the Stevens Institute of Technology, Hoboken, NJ 07030]

Practice Set 1.2

10. A newlywed couple wants to rent a taxicab from their hotel to downtown Chicago. If a taxicab charges $3 upon a customer's entering the taxi, then $0.25 for each $\frac{1}{4}$ mile traveled, how much will the couple pay for a 12-mile taxi ride?

10. _____

11. The bar graph below displays the average hours studied on each day of the week. Use the graph to answer the following questions.
a) On which day(s) of the week do students study the most?
b) On which day(s) of the week do students study the least?
c) How many total hours per week does a student spend studying?

11. a) _____

b) _____

c) _____

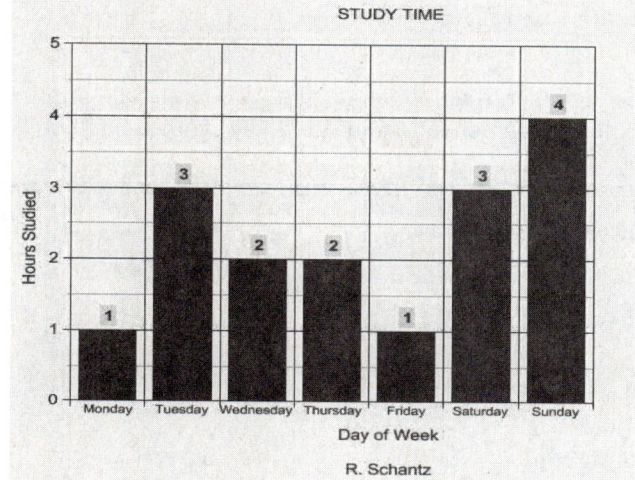

12. The graph below compares the Withdrawal Symptom Scores of women (top line) and men (bottom line) of former smokers who have begun nicotine replacement therapies, such as nicotine patches or nicotine gum. The greater the Withdrawal Symptom Score, the greater the cigarette cravings, irritability, anxiety and/or excessive hunger. Use the graph to answer the questions.
a) Who has the greatest difficulty after quitting smoking, men or women?
b) When do women have the most difficulty after quitting?
c) What is the Withdrawal Symptom Score for the women 4 weeks after quitting?
d) What is the Withdrawal Symptom Score for the men 4 weeks after quitting?

12. a) _____

b) _____

c) _____

d) _____

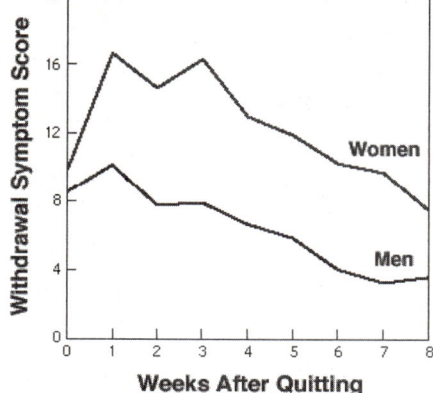

[Source: National Institute of Drug Abuse]

Name:
Instructor:

Date:
Section:

1.3 Fractions

> **Objectives**
> 1. Learn multiplication symbols and recognize factors.
> 2. Prime numbers and greatest common factor.
> 3. Simplify fractions.
> 4. Multiply fractions.
> 5. Divide fractions.
> 6. Add and subtract fractions.
> 7. Change mixed numbers to fractions and vice versa.
>
> **Key Vocabulary**
> variable, factor, prime factorization, prime number, fraction, numerator, denominator, simplify, greatest common factor (GCF), evaluate, least common denominator (LCD), mixed number, whole number

1 Learn multiplication symbols and recognize factors.

Example 1

    a) Show five different ways that "7 times y" can be written.

    b) List all of the positive factors of 24 in ascending order.

2 Prime numbers and greatest common factor.

Example 2 Determine the GCF of the following groups of numbers.

    a) 4, 10   b) 24, 36   c) 12, 21   d) 120, 210

3 Simplify fractions.

Example 3 Simplify.

    a) $\dfrac{4}{10}$ b) $\dfrac{24}{36}$ c) $\dfrac{9}{27}$ d) $\dfrac{120}{210}$

4 Multiply fractions.

Example 4 Find each product. Simplify your answer.

    a) $\dfrac{2}{3} \cdot \dfrac{9}{11}$ b) $\dfrac{3}{11} \cdot \dfrac{2}{9}$ c) $\dfrac{24}{36} \cdot \dfrac{9}{18}$ d) $\dfrac{15}{32} \cdot \dfrac{32}{15}$

Answers: 1a) $7y$, $(7)y$, $(7)(y)$, $7(y)$, $7 \cdot y$ b) 1, 2, 3, 4, 6, 8, 12, 24   2. a) 2   b) 12   c) 3   d) 30   3. a) $\dfrac{2}{5}$ b) $\dfrac{2}{3}$ c) $\dfrac{1}{3}$ d) $\dfrac{4}{7}$

4. a) $\dfrac{6}{11}$ b) $\dfrac{2}{33}$ c) $\dfrac{1}{3}$ d) 1

Examples 1.3

5 **Divide fractions.**

Example 5 Find each quotient. Simplify your answer.

a) $\dfrac{10}{9} \div \dfrac{5}{3}$ b) $\dfrac{4}{7} \div 12$ c) $\dfrac{15}{32} \div \dfrac{15}{32}$ d) $\dfrac{33}{5} \div \dfrac{11}{25}$

6 **Add and subtract fractions.**

Review LCD.

Example 6 Find the LCD of the following pairs of fractions.

a) $\dfrac{22}{25}, \dfrac{7}{25}$ b) $\dfrac{6}{7}, \dfrac{3}{5}$ c) $\dfrac{7}{16}, \dfrac{5}{12}$ d) $\dfrac{1}{9}, \dfrac{2}{3}$

Example 7 Add or subtract. Simplify your answer.

a) $\dfrac{22}{25} - \dfrac{7}{25}$ b) $\dfrac{6}{7} - \dfrac{3}{5}$ c) $\dfrac{7}{16} + \dfrac{5}{12}$ d) $5\dfrac{1}{9} + 3\dfrac{2}{3}$

7 **Change mixed numbers to fractions and vice versa.**

Example 8 Convert the mixed number to a fraction.

a) $5\dfrac{1}{8}$ b) $3\dfrac{2}{3}$

Example 9 Convert the fraction to a mixed number.

a) $\dfrac{33}{5}$ b) $\dfrac{46}{9}$

Answers: 5. a) $\dfrac{2}{3}$ b) $\dfrac{1}{21}$ c) 1 d) 15   6. a) 25 b) 35 c) 48 d) 9   7. a) $\dfrac{3}{5}$ b) $\dfrac{9}{35}$ c) $\dfrac{41}{48}$ d) $8\dfrac{7}{9}$ 8. a) $\dfrac{41}{8}$ b) $\dfrac{11}{3}$
9. a) $6\dfrac{3}{5}$ b) $5\dfrac{1}{9}$

Name:
Instructor:

Date:
Section:

Practice Set 1.3

Use the choices below to fill in each blank.

    variable(s)   factor(s)   fraction   numerator   denominator   simplify

    GCF   LCD   LCM   evaluate   mixed number   solve

1. _____ such as x, y, and z are commonly used in algebra.

2. The numbers 6 and 5 are _____ of the number 30.

3. A _____ contains a numerator and denominator.

4. A fraction can be simplified by dividing both the numerator and denominator by the _____.

5. The _____ of $\frac{3}{5}$ and $\frac{7}{10}$ is 10.

6. $4\frac{7}{10}$ is the _____ for $\frac{47}{10}$.

Determine the GCF of each pair of numbers.

7. 8 and 36   7. _____

8. 105 and 126   8. _____

Simplify the following fractions.

9. $\dfrac{25}{30}$ 10. $\dfrac{54}{72}$ 9. _____

                                                                                       10. _____

Perform the indicated operation. Simplify your answer.

11. $\dfrac{7}{15} - \dfrac{2}{15}$ 12. $\dfrac{7}{15} + \dfrac{1}{30}$ 11. _____

                                                                                         12. _____

13. $\dfrac{7}{9} + \dfrac{3}{8}$ 14. $6\dfrac{3}{5} + 7\dfrac{1}{2}$ 13. _____

                                                                                         14. _____

Practice Set 1.3

15. $\dfrac{5}{8} \cdot \dfrac{10}{25}$

16. $\left(3\dfrac{1}{4}\right)\left(\dfrac{2}{3}\right)$

15. _____

16. _____

17. $\left(\dfrac{14}{25}\right) \div \left(\dfrac{2}{5}\right)$

18. $\dfrac{3}{10} \div 20$

17. _____

18. _____

Convert the fraction to a mixed number or the mixed number to a fraction.

19. $8\dfrac{2}{7}$

20. $\dfrac{540}{360}$

19. _____

20. _____

Problem Solving

21. A recipe that yields $\dfrac{2}{3}$ cup Chermoula, or Moroccan Green Sauce, calls for the following spices:

   $1\dfrac{1}{2}$ teaspoon sweet paprika $\dfrac{1}{2}$ teaspoon ground cumin

   $\dfrac{1}{3}$ cup chopped parsley $\dfrac{1}{8}$ teaspoon cayenne

   $\dfrac{2}{3}$ cup chopped cilantro

   a) Eli wants to triple the recipe for an upcoming dinner party. How much sauce will this amount yield?

   b) Find the exact amounts of each of the listed spices Eli needs to triple the recipe. Simplify your answers.

21. a) _____

   b) _____

   _____

   _____

   _____

   _____

Name:
Instructor:

Date:
Section:

1.4 The Real Number System

Objectives
1. Identify sets of numbers.
2. Know the structure of the real numbers.

Key Vocabulary
set, elements, empty (null) set, counting (natural) numbers/positive integers, integers, rational numbers, irrational numbers, real number, real number line

---

1 Identify sets of numbers.

Example 1 List the following sets of numbers that are being described. If the set doesn't exist, state as such.

a) The set of even counting numbers less than 11

b) The set of whole numbers less than 5

c) The set of negative whole numbers

d) The set of counting numbers that are multiples of 3

2 Know the structure of the real numbers.

Example 2 Consider the following set of numbers.
$$\left\{-3, \frac{3}{5}, 0, 0.\overline{4}, 10, \sqrt{7}, 1.23\right\}$$

List the elements of the set that are

a) natural numbers

b) whole numbers

c) integers

d) rational numbers

e) irrational numbers

f) real numbers

Answers: 1. a) {2, 4, 6, 8, 10} b) {0, 1, 2, 3, 4} c) doesn't exist d) {3, 6, 9, 12, 15, ...} 2. a) {10} b) {0, 10} c) $\left\{-3, 0, 10\right\}$ d) $\left\{-3, \frac{3}{5}, 0, 0.\overline{4}, 10, 1.23\right\}$ e) $\left\{\sqrt{7}\right\}$ f) $\left\{-3, \frac{3}{5}, 0, 0.\overline{4}, 10, \sqrt{7}, 1.23\right\}$

Name:
Instructor:

Date:
Section:

Practice Set 1.4

Match the letter on the right to its best description on the left.

1. The whole number that is not a counting number   _____   **A.** integers

2. The whole numbers combined with the negative integers   _____   **B.** real numbers

3. The integer(s) between -1 and 1 _____   **C.** rational numbers

4. The rational numbers combined with the irrational numbers   _____   **D.** whole numbers

     E. 0

Consider the following set of numbers.

$$\left\{4.3, \frac{9}{11}, -4.3, \sqrt{10}, 0, -\frac{7}{8}, 32\right\}$$

List the set of numbers that are

5. natural numbers   5. _____
6. whole numbers   6. _____
7. integers   7. _____
8. rational numbers   8. _____
9. irrational numbers   9. _____
10. real numbers   10. _____

Determine whether the following statements are true (T) or false (F).

11. $\sqrt{3}$ is a rational number.   11. _____

12. 0 is not a negative integer.   12. _____

13. 0 is not a rational number.   13. _____

14. Every rational number is a real number.   14. _____

15. Not every real number is a counting number.   15. _____

Give an example of a number that satisfies the following conditions if possible.

16. An integer that is not a whole number   16. _____

17. A rational number that is not an integer   17. _____

18. A whole number that is not a counting number   18. _____

19. A whole number that is not an integer   19. _____

20. A negative integer that is not a rational number   20. _____

Name:
Instructor:

Date:
Section:

1.5 Inequalities and Absolute Value

Objectives
1. Determine which is the greater of two numbers.
2. Find the absolute value of a number.

Key Vocabulary
absolute value

Example 1 Translate the following mathematical statements into words.

a) $3 < 5$

b) $-3 > -5$

1 Determine which is the greater of two numbers.

Example 2 Insert either $<$ or $>$ in each shaded area to make a true statement.

a) $5 \; \square \; -5$

b) $-5 \; \square \; -3$

c) $-11 \; \square \; -14$

d) $0.039 \; \square \; 0.05$

e) $0.23 \; \square \; \frac{1}{4}$

f) $-\frac{5}{3} \; \square \; -1.3$

2 Find the absolute value of a number.

Example 3 Evaluate.

a) $|5|$

b) $|-5|$

c) $-|15|$

d) $-|-15|$

Answers: 1a) Three is less than five. b) Negative three is greater than negative five.   2a) > b) < c) > d) < e) < f) <   3a) 5 b) 5 c) –15 d) –15

Name:
Instructor:

Date:
Section:

Practice Set 1.5

Insert either < or > in each shaded area to make a true statement.

1. 11 ☐ 14
2. −11 ☐ 14
3. −11 ☐ −14
4. −14 ☐ −11
5. $\dfrac{1}{2}$ ☐ $\dfrac{1}{8}$
6. $-\dfrac{1}{2}$ ☐ $-\dfrac{1}{8}$
7. $-\dfrac{3}{4}$ ☐ $-\dfrac{5}{8}$
8. 0.04 ☐ 0.1
9. −0.04 ☐ −0.1
10. -0.03 ☐ $-\left(-\dfrac{1}{8}\right)$

Evaluate.

11. $|6|$
12. $|-7|$
13. $-|13|$
14. $-|-7|$

11. _____

12. _____

13. _____

14. _____

Insert either < or > in each shaded area to make a true statement.

15. $|8|$ ☐ $|-4|$
16. $|6|$ ☐ $-|6|$
17. $-|-8|$ ☐ $-|-6|$
18. $\left|-\dfrac{7}{8}\right|$ ☐ $|-0.8|$

15. _____

16. _____

17. _____

18. _____

Challenge

Insert either < or > in each shaded area to make a true statement. Let a, b, and c be real numbers where $a < 0$, $b > 0$ and $c < 0$.

19. a ☐ $|c|$
20. $a \cdot |b|$ ☐ $|a \cdot b|$

Name:
Instructor:

Date:
Section:

1.6 Addition of Real Numbers

> **Objectives**
> 1. Add real numbers using a number line.
> 2. Add fractions.
> 3. Identify opposites.
> 4. Add using absolute values.
>
> **Key Vocabulary**
> operations, addend, sum, additive inverses (opposites)

1 Add real numbers using a number line.

Example 1 Evaluate the following using a number line.

a) $5+(-4)$

b) $-5+(4)$

c) $5+(4)$

d) $-5+(-4)$

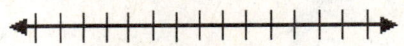

2 Add fractions.

Example 2 Add.

a) $\dfrac{4}{5}+\dfrac{5}{7}$

b) $-\dfrac{4}{5}+\dfrac{5}{7}$

c) $\dfrac{4}{5}+\left(-\dfrac{5}{7}\right)$

d) $-\dfrac{4}{5}+\left(-\dfrac{5}{7}\right)$

3 Identify opposites.

Example 3 Write the opposite of each number.

a) 8

b) -8

c) $\dfrac{5}{8}$

d) $-\dfrac{5}{8}$

4 Add using absolute values.

Example 4 Add the following numbers using absolute values.

a) $14+(-12)$

b) $(-14)+(-12)$

c) $\left(-\dfrac{5}{8}\right)+\left(\dfrac{1}{4}\right)$

d) $1.25+(0.20)$

Answers: 1a) 1 b) -1 c) 9 d) -9 2a) $\dfrac{53}{35}$ or $1\dfrac{18}{35}$ b) $-\dfrac{3}{35}$ c) $\dfrac{3}{35}$ d) $-\dfrac{53}{35}$ or $-1\dfrac{18}{35}$ 3a) -8 b) 8 c) $-\dfrac{5}{8}$ d) $\dfrac{5}{8}$ 4a) 2 b) -26 c) $-\dfrac{3}{8}$ d) 1.45

Name:
Instructor:

Date:
Section:

Practice Set 1.6

Use the choices below to fill in each blank.

positive   negative   additive inverse   opposite(s)
reverse   operation(s)   absolute value(s)   zero

1. The sum of two positive numbers is a _____ number.

2. The sum of two negative numbers is a _____ number.

3. The sum of a number and its _____ or _____ will always be 0.

4. When adding numbers that have the same sign, we add their _____ .

5. Since the sum of 12 and -12 is 0, -12 is the _____ or _____ of 12.

6. Addition, subtraction, multiplication, and division are four basic _____ of arithmetic.

For each of the following, a) determine by observation whether the sum will be a positive number, zero, or a negative number, and b) find each sum.

7. $(-8.02)+(8.10)$ 7. a) _____

                                                                                               b) _____

8. $(-548)+(548)$ 8. a) _____

                                                                                               b) _____

9. $\dfrac{1}{3}+\left(-\dfrac{1}{2}\right)$ 9. a) _____

                                                                                               b) _____

Add.

10. $6+(-12)$ 10. _____

11. $-11+(-4)$ 11. _____

12. $-12+(6)$ 12. _____

Practice Set 1.6

13. $30.4 + (-31)$ 13. _____

14. $-78 + (-10)$ 14. _____

15. $-12 + (-6)$ 15. _____

16. $0 + (-7)$ 16. _____

17. $-78 + (10)$ 17. _____

18. $-12 + (12)$ 18. _____

19. $-\dfrac{11}{12} + \left(-\dfrac{5}{6}\right)$ 19. _____

20. $\dfrac{-7}{9} + \left(\dfrac{7}{9}\right)$ 20. _____

21. $\dfrac{15}{24} + \left(-\dfrac{9}{16}\right)$ 21. _____

Problem Solving

22. Overdrawn Checking Account Mr. Knight was unaware that his checking account was overdrawn by $23. If he then wrote a check for $94, find Mr. Knight's overdrawn balance.   22. _____

Name:
Instructor:

Date:
Section:

1.7 Subtraction of Real Numbers

Objectives
1. **Subtract numbers.**
2. **Subtract numbers mentally.**
3. **Evaluate expressions containing more than two numbers.**

Key Vocabulary
difference

1 Subtract numbers.

Example 1 Evaluate.

a) $10-(+3)$ b) $10-(-3)$ c) $-10-3$

d) $-10-(-10)$ e) $-\dfrac{4}{9}-\dfrac{3}{8}$ f) $-14.6-15.4$

2 Subtract numbers mentally.

Example 2 Subtract mentally.

a) $10-14$ b) $-10-14$ c) $-10-1.4$

d) $-\dfrac{5}{11}-\dfrac{13}{22}$ e) $10-1.4$ f) $\dfrac{1}{2}-\dfrac{3}{4}$

3 Evaluate expressions containing more than two numbers.

Example 3 Evaluate.

a) $-12-(+3)+5$ b) $-12-(-3)-(-5)$

c) $40-25+15-10$ d) $1-2+3-4+5-6$

Answers: 1a) 7 b) 13 c) -13 d) 0 e) $-\dfrac{59}{72}$ f) -30 2a) -4 b) -24 c) -11.4 d) $-\dfrac{23}{22}$ or $-1\dfrac{1}{22}$ e) 8.6 f) $-\dfrac{1}{4}$
3a) -10 b) -4 c) 20 d) -3

Name:
Instructor:

Date:
Section:

Practice Set 1.7

Evaluate.

1. $1-2$

2. $-8-8$

3. $-8-(-8)$

4. $8-8$

5. $-31-11$

6. $0-0.9$

7. $31-(-11)$

8. $0-(-0.9)$

9. Subtract 18 from 5.

10. Subtract -4.9 from -1.1.

11. $\dfrac{5}{12} - \dfrac{3}{10}$

12. $\dfrac{3}{10} - \left(-\dfrac{5}{12}\right)$

13. $\dfrac{3}{16} - \dfrac{7}{8}$

14. $\dfrac{3}{16} - \left(-\dfrac{7}{8}\right)$

15. $-\dfrac{4}{7} - \dfrac{1}{3}$

16. $\dfrac{2}{3} - \left(-\dfrac{3}{4}\right)$

17. Subtract $\dfrac{3}{5}$ from $\dfrac{2}{7}$.

18. Subtract $-\dfrac{5}{7}$ from $\dfrac{4}{5}$.

19. $-8-2+7-(-3)$

20. $-9-20+(13)-(-10)$

1. _____
2. _____
3. _____
4. _____
5. _____
6. _____
7. _____
8. _____
9. _____
10. _____
11. _____
12. _____
13. _____
14. _____
15. _____
16. _____
17. _____
18. _____
19. _____
20. _____

Name:
Instructor:
Date:
Section:

1.8 Multiplication and Division of Real Numbers

Objectives
1. Multiply numbers.
2. Divide numbers.
3. Remove negative signs from denominators.
4. Evaluate divisions involving 0.

Key Vocabulary
like/unlike signs, positive/negative numbers, product, quotient, undefined

1 Multiply numbers.

Example 1 Evaluate.

a) $10(-3)$
b) $-10(3)$
c) $-10(-3)$
d) $(-12)(0)(-2)$

e) $(-4)(-2)(1)(-3)$
f) $-8(0)$
g) $\left(\dfrac{11}{15}\right)\left(\dfrac{-5}{22}\right)$
h) $\left(\dfrac{-9}{8}\right)\left(\dfrac{-8}{9}\right)$

2 Divide numbers.

Example 2 Evaluate.

a) $\dfrac{-54}{9}$
b) $\dfrac{-32}{8}$
c) $\dfrac{-2.8}{0.2}$

d) $\dfrac{0}{-8}$
e) $\dfrac{-24}{-3}$
f) $\dfrac{-20}{20}$

3 Remove negative signs from denominators.

Example 3 Write the fraction $\dfrac{8}{-9}$ two other ways.

4 Evaluate divisions involving 0.

Example 4 Indicate whether each quotient is 0 or undefined.

a) $\dfrac{0}{20}$
b) $\dfrac{20}{0}$
c) $0 \div 9$
d) $-10 \div 0$

Answers: 1a) -30 b) -30 c) 30 d) 0 e) -24 f) 0 g) $-\dfrac{1}{6}$ h) 1 2a) -6 b) -4 c) -14 d) 0 e) 8 f) -1
3. $\dfrac{-8}{9}, -\dfrac{8}{9}$ 4a) 0 b) undefined c) 0 d) undefined

Name:
Instructor:

Date:
Section:

Practice Set 1.8

Use the choices below to fill in each blank.

positive negative undefined opposite(s)
reverse operation same zero

1. The product of any real number and _____ is 0.

2. The product of two negative numbers is a _____ number.

3. The quotient of two negative numbers is a _____ number.

4. The quotient of two numbers that have _____ signs is negative.

5. The product of an even number of negative numbers is a _____ number.

6. Zero divided by a negative number is _____ .

Find each product.

7. $(-9)(3)$

8. $(-9)(-3)$

9. $(6)(-8)$

10. $(-7)(3)(-2)$

11. $(-1)(1)(-1)(1)$

12. $(-9)(0)(3)$

13. $(-11.2)(10)$

14. $\left(-\dfrac{2}{3}\right)\left(-\dfrac{3}{2}\right)$

15. $\left(\dfrac{6}{7}\right)\left(-\dfrac{14}{15}\right)$

16. $\left(\dfrac{-11}{12}\right)\left(-\dfrac{5}{6}\right)$

17. $\left(\dfrac{-7}{9}\right)\left(\dfrac{7}{9}\right)$

18. $\left(\dfrac{15}{24}\right)\left(\dfrac{9}{-16}\right)$

7. _____
8. _____
9. _____
10. _____
11. _____
12. _____
13. _____
14. _____
15. _____
16. _____
17. _____
18. _____

20 Angel/Runde, *Elementary Algebra for College Students*, 9e Copyright © 2015 by Pearson Education, Inc.

Practice Set 1.8

Find each quotient.

19. $(-120) \div (-12)$

20. $\dfrac{28}{-49}$

19. _____

20. _____

21. $\dfrac{-42}{-7}$

22. $\left(\dfrac{-24}{9}\right) \div \left(\dfrac{-16}{3}\right)$

21. _____

22. _____

23. $\left(\dfrac{-2}{3}\right) \div \left(\dfrac{-3}{2}\right)$

24. $-18 \div \left(\dfrac{-12}{5}\right)$

23. _____

24. _____

25. $\dfrac{63.5}{-10}$

26. $\dfrac{-42}{0}$

25. _____

26. _____

27. $\dfrac{0}{-5}$

28. Divide 24 by 0.

27. _____

28. _____

Problem Solving

29. Weather Forecast The meteorologist on T.V. tonight made the following prediction: "Folks, tonight the temperature will drop down to $-3°$ F. If you think *that's* cold, tomorrow night will be triple tonight's temperature." How cold will it be tomorrow night?

29. _____

Challenge

30. $(-1)^{100} + (-1)^{101}$

30. _____

Name:
Instructor:
Date:
Section:

1.9 Exponents, Parentheses, and the Order of Operations

> **Objectives**
> 1. Learn the meaning of exponents.
> 2. Evaluate expressions containing exponents.
> 3. Learn the difference between $-x^2$ and $(-x)^2$.
> 4. Learn the order of operations.
> 5. Learn the use of parentheses.
> 6. Evaluate expressions containing variables.
>
> **Key Vocabulary**
> base, exponent, order of operations, parentheses ( ), PEMDAS, nested grouping symbols

1 Learn the meaning of exponents.

Example 1 Complete the following chart by identifying the base and exponent of each expression.

	Expression	Base	Exponent
a)	3^4		
b)	$(-3)^4$		
c)	$(x+3)^2$		
d)	-3^4		

2 Evaluate expressions containing exponents.

Example 2 Evaluate the following expressions.

a) 3^4 b) $(-3)^4$ c) $\left(-\dfrac{1}{3}\right)^3$ d) $\left(\dfrac{3}{4}\right)^2$

3 Learn the difference between $-x^2$ and $(-x)^2$.

Example 3 Evaluate the following expressions.

a) -3^4 b) $(-2)^2$ c) -2^2

d) $\left(-\dfrac{3}{4}\right)^2$ e) -6^2 f) $(-6)^2$

Answers: 1a) 3; 4 b) -3; 4 c) $(x+3)$; 2 d) 3; 4 2a) 81 b) 81 c) $-\dfrac{1}{27}$ d) $\dfrac{9}{16}$ 3a) -81 b) 4 c) -4 d) $\dfrac{9}{16}$ e) -36 f) 36

Examples 1.9

4 Learn the order of operations.

Example 4 Next to each step, identify the operation used to get to that step (place a P, E, M, D, A, or S in each box).

a) $\quad 6 - 4 \cdot 3^2 + 2^3$

$\quad\quad 6 - 4 \cdot 9 + 8 \quad \underline{}$

$\quad\quad 6 - 36 + 8 \quad \underline{}$

$\quad\quad 14 - 36 \quad \underline{}$

$\quad\quad -22 \quad \underline{}$

b) $\quad 10(3-5)^2 - 12 \div 3 \cdot 2$

$\quad\quad 10(-2)^2 - 12 \div 3 \cdot 2 \quad \underline{}$

$\quad\quad 10(4) - 12 \div 3 \cdot 2 \quad \underline{}$

$\quad\quad 40 - 12 \div 3 \cdot 2 \quad \underline{}$

$\quad\quad 40 - 4 \cdot 2 \quad \underline{}$

$\quad\quad 40 - 8 \quad \underline{}$

$\quad\quad 32 \quad \underline{}$

Example 5 Use the order of operations to simplify the following expressions.

a) $\;\; -7 - 3^2 + 6 \cdot 2 \quad$ b) $\;\; 4\left(\dfrac{1}{2}\right)^3 - 2\left(\dfrac{1}{3}\right) \quad$ c) $\;\; 10(4) - 12 \div (3 \cdot 2) \quad$ d) $\;\; -2(-3+1)^2 - 8 \div 2 \cdot 5$

5 Learn the use of parentheses.

Example 6 Evaluate the following expressions.

a) $\;\; 2[3-(6-7)]^2 \quad$ b) $\;\; -3\{[(8+1)-6]+6\} \quad$ c) $\;\; 1-\left[1-(1-2)^2\right] \quad$ d) $\;\; -3\{6+[6-(1+8)]\}$

6 Evaluate expressions containing variables.

Example 7 Evaluate $2x^2 + 4xy - 1$ for the given values of the variables.

a) $\;\; x = -3, y = 2 \quad$ b) $\;\; x = 3, y = -2 \quad$ c) $\;\; x = \dfrac{1}{3}, y = 3 \quad$ d) $\;\; x = -2, y = 0$

Answers: 4a) E; M; A; S b) P; E; M; D; M; S   5a) –4 b) $\dfrac{-1}{6}$ c) 38 d) –28   6a) 32 b) –27 c) 1 d) –9   7a) –7 b) –7 c) $\dfrac{29}{9}$ d) 7

Copyright © 2015 by Pearson Education, Inc.   Angel/Runde, *Elementary Algebra for College Students*, 9e   23

Name:
Instructor:

Date:
Section:

Practice Set 1.9

Evaluate.

1. 8^2

2. $\left(\dfrac{1}{8}\right)^2$

3. 4^3

4. $\left(\dfrac{2}{3}\right)^3$

1. _____

2. _____

3. _____

4. _____

For the following, a) determine whether the answer should be positive (+) or negative (−), and b) evaluate the expression by observation.

5. $(-3)^2$

5. a) _____

b) _____

6. -3^2

6. a) _____

b) _____

7. $(-1)^{10}$

7. a) _____

b) _____

8. -3^3

8. a) _____

b) _____

9. $(-3)^3$

9. a) _____

b) _____

10. $-(-3)^2$

10. a) _____

b) _____

11. $(-1)^{11}$

11. a) _____

b) _____

12. $-(-3)^3$

12. a) _____

b) _____

Practice Set 1.9

Evaluate.

13. $5+(12-4\cdot 2)^2$

14. $3-4^2+2\cdot 3$

15. $-3^2+2(9)\left(\dfrac{1}{3}\right)^3$

16. $5(8-10\cdot 2)\div[(-5)(6)]$

17. $-15-[-3(7)\div(-1)(-3)]\div 3$

18. $12\div(2)(3)-6^2\div 2$

19. $\left(\dfrac{2}{9}-\dfrac{3}{4}\right)\cdot 6^2-1$

20. $\dfrac{-(5-6)^2+4}{-5(-1)^4-2^2}$

13. _____

14. _____

15. _____

16. _____

17. _____

18. _____

19. _____

20. _____

Evaluate each expression for the given value of the variable or variables.

21. $9(x-2);\ x=11$

22. $4(x-y)+7x;\ x=1,\ y=0$

23. $x^2-y^2;\ x=-1,\ y=\dfrac{1}{2}$

24. $9x^2-12xy+4y^2;\ x=-4,\ y=-6$

21. _____

22. _____

23. _____

24. _____

Name:
Instructor:
Date:
Section:

1.10 Properties of the Real Number System

> **Objectives**
> 1. Learn the commutative property.
> 2. Learn the associative property.
> 3. Learn the distributive property.
> 4. Learn the identity properties.
> 5. Learn the inverse properties.
>
> **Key Vocabulary**
> commutative property of addition/multiplication, associative property of addition/multiplication, distributive property, identity element (additive/multiplicative identity), additive inverse, multiplicative inverse

1 Learn the commutative property.

Example 1 Use the commutative property of addition to complete each statement.

a) $8 + 4 =$ _____   b) $7 + x =$ _____

Example 2 Use the commutative property of multiplication to complete each statement.

a) $4(3) =$ _____   b) $4(x + 2) =$ _____

2 Learn the associative property.

Example 3 Use the associative property of addition to complete each statement.

a) $8 + (4 + 2) =$ _____   b) $(7 + x) + 2 =$ _____

Example 4 Use the associative property of multiplication to complete each statement.

a) $(8 \cdot 4) \cdot 2 =$ _____   b) $7 \cdot (2 \cdot x) =$ _____

3 Learn the distributive property.

Example 5 Use the distributive property to complete each statement.

a) $2(7 + 10) =$ _____   b) $-2x - 2y =$ _____

Answers: 1a) $4 + 8$ b) $x + 7$ 2a) $3(4)$ b) $(x + 2) \cdot 4$ 3a) $(8 + 4) + 2$ b) $7 + (x + 2)$ 4a) $8 \cdot (4 \cdot 2)$ b) $(7 \cdot 2) \cdot x$
5a) $2 \cdot 7 + 2 \cdot 10$ b) $-2(x + y)$

Examples 1.10

4 Learn the identity properties.

Example 6 Use the appropriate identity property to complete each statement.

a) $(\underline{}) + 0 = 7$

b) $(\underline{}) \cdot 1 = 7$

c) $(\underline{}) + 0 = \dfrac{1}{7}$

d) $(\underline{}) \cdot 1 = -\dfrac{1}{7}$

e) $(\underline{}) + 0 = y$

f) $(\underline{}) \cdot 1 = (a+b)$

5 Learn the inverse properties.

Example 7 Determine the (1) additive inverse, and (2) multiplicative inverse for each expression.

a) 9

b) $\dfrac{1}{9}$

c) y

d) $-\dfrac{1}{9}$

e) $\dfrac{2}{3}$

f) $-y$

Example 8 Use the appropriate inverse property to complete each statement.

a) $4 + (\underline{}) = 0$

b) $9 \cdot (\underline{}) = 1$

c) $-y + (\underline{}) = 0$

d) $-\dfrac{1}{9} \cdot (\underline{}) = 1$

Answers: 6a) 7 b) 7 c) $\dfrac{1}{7}$ d) $-\dfrac{1}{7}$ e) y f) $(a+b)$ 7a) (1) -9; (2) $\dfrac{1}{9}$ b) (1) $-\dfrac{1}{9}$; (2) 9 c) (1) $-y$; (2) $\dfrac{1}{y}$ d) (1) $\dfrac{1}{9}$; (2) -9 e) (1) $-\dfrac{2}{3}$; (2) $\dfrac{3}{2}$ f) (1) y; (2) $-\dfrac{1}{y}$ 8a) -4 b) $\dfrac{1}{9}$ c) y d) -9

Name:
Instructor:
Date:
Section:

Practice Set 1.10

Use the choices below to fill in each blank.

$$0 \qquad -1 \qquad -\frac{1}{3} \qquad \frac{1}{0}$$

$$1 \qquad \frac{1}{3} \qquad -3 \qquad 3$$

1. The identity element of multiplication is _____.
2. The identity element of addition is _____.
3. The multiplicative inverse of $\frac{1}{3}$ is _____.
4. The additive inverse of 3 is _____.

Name each indicated property.

5. $(3)+(-9)=(-9)+(3)$
6. $(-3)(1)=-3$

7. $(-9)(3)=(3)(-9)$
8. $7+x=x+7$

9. $-7 \cdot \frac{-1}{7}=1$
10. $-7+0=-7$

11. $7+(x+2)=7+(2+x)$
12. $\frac{1}{2}+\left(-\frac{1}{2}\right)=0$

13. $7 \cdot (x \cdot 2)=7 \cdot (2 \cdot x)$
14. $\left(\frac{1}{2} \cdot 9\right) \cdot (8)=\frac{1}{2} \cdot (9 \cdot 8)$

15. $5(x+2)=5x+10$
16. $\frac{-7}{9}+\left(\frac{7}{9}\right)=0$

17. $3(10-4)=30-12$
18. $\left(\frac{-2}{3}\right)\left(\frac{3}{-2}\right)=1$

19. $6-3y=3(2-y)$
20. $-7 \cdot 1=-7$

5. _____
6. _____
7. _____
8. _____
9. _____
10. _____
11. _____
12. _____
13. _____
14. _____
15. _____
16. _____
17. _____
18. _____
19. _____
20. _____

Chapter 1 Vocabulary Reference Sheet

Term	Definition	Example
five-step problem-solving procedure	Guidelines for solving word problems developed by George Polya	Steps are: Understand, Translate, Carry out calculations, Check, and Answer the question.
(algebraic) expression	A collection of numbers or letters, grouping symbols (such as parentheses), and operations	$5(18 \div 3)$
operations	Addition, subtraction, multiplication, and division	$2 + 3 - 7 \div 5 \cdot 3$
bar graph	A way to display data using bars	
line graph	A way to display data, usually over time, using a line	
circle graph	A way to display data that represent parts of a whole	
measures of central tendency (or averages)	Values that are representative of a set of data	Mean and median
mean	Adding all the values and dividing the sum by the number of values	$\frac{86 + 92 + 76 + 88}{4} = 85.5$
median	The value in the middle of a set of ranked data	Given the set of temperatures $\{76, 86, 88, 92\}$, median temperature is 87.
ranked data	A list of data in numerical order	The data $\{76, 86, 88, 92\}$ are ranked.
variable	A letter or symbol used to represent an unknown quantity	$x, y, z, a, b, c, X, Y, \blacksquare, \Delta$
factor	Numbers multiplied together	In $2x$, 2 and x are factors.
fraction	A way to express parts of a whole	$\frac{2}{5}$ is a fraction representing 2 parts of 5.
numerator	The top number of a fraction	In $\frac{2}{5}$, the numerator is 2.
denominator	The bottom number of a fraction	In $\frac{2}{5}$, the denominator is 5.
simplify a fraction	To write an expression in lowest terms	$\frac{14}{35}$ simplified is $\frac{2}{5}$.
prime number	A natural number divisible only by itself and 1	2, 3, 5, and 7 are the prime numbers less than 10.
prime factorization	Writing a number as a product of prime numbers	$66 = 2 \cdot 3 \cdot 11$

Chapter 1 Vocabulary

greatest common factor (GCF)	The largest number that will divide both the numerator and denominator	The GCF of 12 and 18 is 6.
evaluate	To find the value of an expression	Evaluate $2x - y$ if $x = 3$ and $y = -1$. Solution: $2(3) - (-1) = 6 + 1 = 7$
least common denominator (LCD)	The smallest number that is divisible by two or more denominators	The LCD of 9 and 12 is 36.
mixed number	A whole number added to a fraction	$2\frac{3}{5}$
whole number	The collection of 0 with the counting numbers, $\{0, 1, 2, 3, 4, 5,...\}$	28
set	Collection of objects	$A = \{1, 2, 3, 4, 5\}$
elements	Objects of a set	1, 2, 3, 4, and 5 are the elements of A.
empty (null) set	A set with no elements, $\{\}$ or $\emptyset$	The set of numbers which are both rational and irrational is the empty set.
natural (counting) numbers	$\{1, 2, 3, 4, 5,...\}$	1
integers	$\{..., -3, -2, -1, 0, 1, 2, 3, ...\}$	-7
rational numbers	Numbers that can be expressed as a quotient of two integers; denominator cannot be zero	$\frac{2}{3}$
irrational numbers	Numbers represented on the number line that are not rational numbers	$\sqrt{5}$
real number	Any rational or irrational number	$1, 0, -15, \frac{2}{3}, \sqrt{5}$
real number line	Graph representing real numbers	
absolute value	The distance between a number and 0 on the number line	$|-3| = 3$
addend	The number being added	In $-8 + 8$, -8 and 8 are addends.
sum	The answer to an addition problem	In $2 + 3 = 5$, 5 is the sum.
additive inverses (opposites)	Any two numbers whose sum is 0	-8 and 8 are additive inverses.
difference	The answer to a subtraction problem	In $3 - 2 = 1$, 1 is the difference.
product	The answer to a multiplication problem	In $2(3) = 6$, 6 is the product.
quotient	The answer to a division problem	In $12 \div 2 = 6$, 6 is the quotient.
undefined	Division by zero	$12 \div 0$ is undefined.
base	A number that is raised to a power	In 2^3, 2 is the base.
exponent	The power of the base	In 2^3, 3 is the exponent.
order of operations	PEDMAS is an acronym for the order of operations, or the order in which operations are to be done: parentheses, exponents, multiplication, division, addition, subtraction	$2 - 7(3-4)^2 = 2 - 7(-1)^2$ (P) $= 2 - 7(1)$ (E) $= 2 - 7$ (M) $= -5$ (S)
nested grouping symbols	Grouping symbols within grouping symbols	$2 - 7\{3 - (-8 - 1) \div 3\}^2$
identity element	A number that when added/multiplied to a real number, the value of the real number doesn't change	Identity element of addition is 0: $2 + 0 = 2$ Identity element of multiplication is 1: $2(1) = 2$
multiplicative inverse	Two numbers whose product is 1	$\frac{1}{8}$ and 8 are multiplicative inverses.

Name:
Instructor:

Date:
Section:

Chapter 1 Practice Test A

Use Polya's five-step problem solving procedure to solve the following problems.

1. Wes Kinworthy purchased a new computer for $749.99, the price before tax. If the sales tax rate is $7\frac{3}{4}$%, find the price of the computer including tax.

   1._____

2. Kim Stegton is considering purchasing a new car with the list price of $15,000. The car dealership is offering a "customer choice" special. The customer can choose either a 4.9% discount or a $500 rebate. Which is the better deal and why?

   2._____
   _____

3. The AADT, or Average Annual Daily Traffic, is a measure of average daily traffic at specific locations. The AADT at Highway 47 and Highway 61 in 1970 was 1700 cars per day. The AADT at the same location 34 years later in 2004 increased to 310 more than seven times the 1970 AADT figure. What was the AADT at this location in 2004? [Source: Missouri Department of Transportation]

   3._____

4. The graph below displays the percentages, by class, of those students who attended summer school at Sunny High School in 2009. In that year, 160 students attended summer school at Sunny High School. Use this information to answer the following questions.
   a) How many freshmen attended summer school?
   b) How many sophomores attended summer school?
   c) How many more juniors than seniors attended summer school?

   4. a)_____

   b)_____

   c)_____

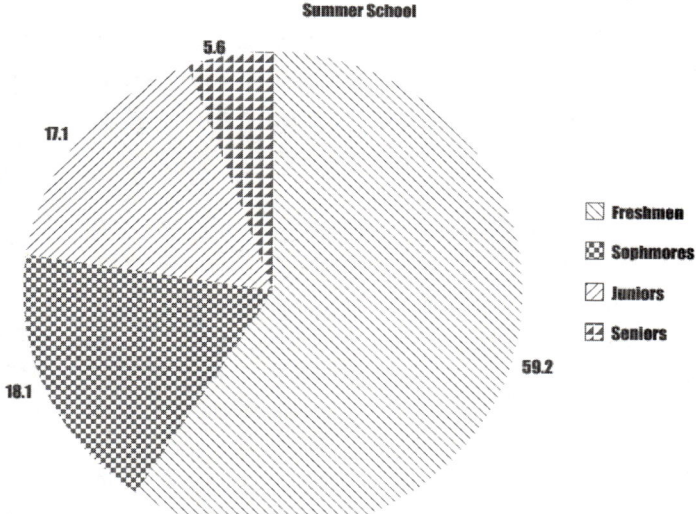

5. Given the following set of numbers, list those that are the rational numbers.
$$\left\{-\frac{1}{2}, \frac{2}{3}, \sqrt{15}, -5, 0, 98, 2.5\right\}$$

   5._____

Copyright © 2015 by Pearson Education, Inc.   Angel/Runde, *Elementary Algebra for College Students*, 9e

Practice Test 1A

6. Jason's weekly gasoline purchases for one month were $34.58, $12.67, $23.01, and $35.82. Calculate the median of Jason's gasoline purchases.

6. _____

Insert either <, >, or = in each blank to make a true statement.

7. $-\dfrac{4}{10}$ _____ -0.09

8. $-|-5|$ _____ $|-5|$

7. _____

8. _____

Evaluate.

9. $\left(\dfrac{10}{27}\right)\left(-\dfrac{9}{30}\right)$

10. $\left(\dfrac{-15}{16}\right) \div \left(\dfrac{25}{-24}\right)$

9. _____

10. _____

11. $\dfrac{3}{5} + \left(-\dfrac{5}{12}\right)$

12. $\dfrac{11}{15} - \dfrac{25}{24}$

11. _____

12. _____

13. $15 - 25 - (-2)$

14. $(-4)(10)(-0.3)(-1)$

13. _____

14. _____

15. $\left(-\dfrac{1}{3}\right)^4$

16. $2 - [(3-5)^3 \div 4 \cdot (-3)]$

15. _____

16. _____

Evaluate the expression for the given values.

17. $7x - 14xy - y^2$; $x = -2, y = 3$

17. _____

18. $-x^2 + x + y$, $x = -1, y = -4$

18. _____

Name each indicated property.

19. $4(x+2) = 4x + 8$

19. _____

20. $(x+5) + 3 = x + (5+3)$

20. _____

32   Angel/Runde, Elementary Algebra for College Students, 9e   Copyright © 2015 by Pearson Education, Inc.

Name:
Instructor:
Date:
Section:

Chapter 1 Practice Test B

1. While on vacation, Patty purchased one handmade doll for $25.30 and five postcards for $1.50 each. If there is a 7.25% sales tax, what was Patty's total bill including tax?
   a) $1.94
   b) $2.38
   c) $28.74
   d) $35.18

2. The line graph below displays the highest daily temperatures over a 5-day period. Which day had the lowest temperature and what was the temperature?

   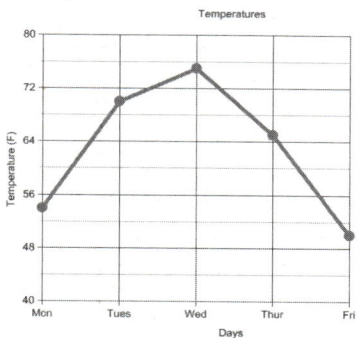

   a) Monday; 54°
   b) Monday; 56°
   c) Friday; 48°
   d) Friday; 50°

3. Kristen, a college junior, is in her 7th semester of college, including summer school. Her course loads for the first 6 semesters of college have been as follows: 15, 13, 6, 12, 15, and 3 credit hours. Choose the expression that will find the mean number of credit hours for her first 6 semesters of college.
   a) $\dfrac{15+13+6+12+15+3}{6}$
   b) $\dfrac{15+13+6+12+15+3}{7}$
   c) $\dfrac{12+13}{6}$
   d) $\dfrac{12+13}{2}$

4. Given the following set of numbers, list those that are integers.
$$\left\{\dfrac{3}{8},\ -6,\ -\dfrac{1}{4},\ 0,\ 45,\ 0.\overline{3},\ \sqrt{21}\right\}$$
   a) $\left\{-6, -\dfrac{1}{4}, 0, 45\right\}$
   b) $\{-6, 0, 45\}$
   c) $\{-6, 45\}$
   d) $\left\{-6, -\dfrac{1}{4}, 45\right\}$

Insert either <, >, or = in each blank to make a true statement.

5. -17 _____ -19
   a) <
   b) >
   c) =

6. $|-0.09|$ _____ $|-0.4|$
   a) <
   b) >
   c) =

Evaluate.

7. $-12+(-16)$
   a) -4
   b) 4
   c) -28
   d) 28

8. $-10-4$
   a) -6
   b) -14
   c) 14
   d) 40

9. $-9-(-11)+6-16$
   a) -30
   b) -8
   c) 8
   d) 24

Practice Test 1B

10. $\left(-\dfrac{1}{3}\right)^2$

   a) $-\dfrac{1}{6}$ b) $\dfrac{1}{6}$ c) $-\dfrac{1}{9}$ d) $\dfrac{1}{9}$

11. $\left(-\dfrac{5}{8}\right)+\left(-\dfrac{2}{5}\right)$

   a) $-\dfrac{1}{4}$ b) $\dfrac{1}{4}$ c) $-\dfrac{41}{40}$ d) $\dfrac{41}{40}$

12. $-\dfrac{4}{7}-\dfrac{1}{2}$

   a) $\dfrac{1}{14}$ b) $-\dfrac{15}{14}$ c) $-\dfrac{1}{14}$ d) $\dfrac{15}{14}$

13. $\left(-\dfrac{15}{8}\right)\left(-\dfrac{2}{3}\right)$

   a) $\dfrac{16}{45}$ b) $-\dfrac{16}{45}$ c) $-\dfrac{5}{4}$ d) $\dfrac{5}{4}$

14. $\left(\dfrac{20}{3}\right)\div\left(-\dfrac{3}{20}\right)$

   a) $-\dfrac{9}{400}$ b) $-\dfrac{400}{9}$ c) 1 d) -1

15. -10^2
   a) 100 b) -100 c) 20 d) -20

16. $-11-(1-4)-5(2)^2$
   a) -108 b) -34 c) -12 d) -28

17. Choose the fraction that is *not* equivalent to $-\dfrac{3}{4}$.

   a) $\dfrac{-3}{4}$ b) $\dfrac{3}{-4}$ c) $-\left(\dfrac{-3}{-4}\right)$ d) $-\left(\dfrac{-3}{4}\right)$

18. Evaluate $-x^2-xy+2y^2$; $x=-3, y=-2$
   a) 19 b) -7 c) -11 d) -19

19. Use the distributive property to complete the statement: $3(x-5)=$
   a) $3(5-x)$ b) $(x-5)\cdot 3$ c) $(3x)-5=$ d) $3x-15$

20. Use the commutative property of addition to complete the statement: $6+(x+10)=$
   a) $6+(10+x)$ b) $(6+10)+x$ c) $6x+60$ d) $6(x+10)$

Name:
Instructor:
Date:
Section:

2.1 Combining Like Terms

> **Objectives**
> 1. Identify terms.
> 2. Identify like terms.
> 3. Combine like terms.
> 4. Use the distributive property.
> 5. Remove parentheses when they are preceded by a plus or minus sign.
> 6. Simplify an expression.
>
> **Key Vocabulary**
> (algebraic) expression, terms, (numerical) coefficient, constant (term), like terms, combine like terms, simplify an expression

1 Identify terms.

Example 1 List the terms of each expression.

a) $6x - 4 - 2y$
b) $9x^2 - x + y - \frac{1}{2}$
c) $1 + 2(x+3) - 4x$
d) $\frac{y-2}{3} + 7x - 1$

Example 2 List the coefficient of each term.

a) $6x$
b) $-y$
c) $\frac{5x}{3}$
d) $\frac{1}{4}x$
e) -3
f) $\frac{x}{2}$
g) 3
h) $\frac{y-2}{4}$

2 Identify like terms.

Example 3 Identify any like terms in each expression.

a) $6x - 4 - 2y + 2x$
b) $3x^2 + 4 + x^2 - 1 - x$
c) $1 - 3x + 5 + \frac{x}{2}$
d) $7x^2 - 7x + x^2 - x$

3 Combine like terms.

Example 4 Combine like terms.

a) $6x - 4 - 2y + 2x$
b) $3x^2 + 4 + x^2 - 1 - x$
c) $1 - 3x + 5 + \frac{x}{2}$
d) $7x^2 - 7x + x^2 - x$

Answers: 1a) $6x$; -4, $-2y$ b) $9x^2$, $-x$, y, $-\frac{1}{2}$ c) $1, 2(x+3), -4x$ d) $\frac{y-2}{3}$, $7x$, -1 2a) 6 b) -1 c) $\frac{5}{3}$ d) $\frac{1}{4}$ e) -3 f) $\frac{1}{2}$ g) 3 h) $\frac{1}{4}$ 3a) $6x, 2x$ b) $3x^2, x^2$; $4, -1$ c) $1, 5$; $-3x, \frac{x}{2}$ d) $7x^2, x^2$; $-7x, -x$ 4a) $8x - 2y - 4$ b) $4x^2 - x + 3$ c) $6 - \frac{5}{2}x$ d) $8x^2 - 8x$

Examples 2.1

4 Use the distributive property.

Example 5 Use the distributive property to remove parentheses.

a) $2(3x-4)$ b) $-3(x+4)$ c) $-5(9-y)$

d) $\frac{2}{3}(3x-6y)$ e) $2(3a-4b+2)$ f) $-\frac{1}{2}(3x-4y+2)$

5 Remove parentheses when they are preceded by a plus or minus sign.

Example 6 Use the distributive property to remove parentheses.

a) $(3x-4)$ b) $-(3x-4)$ c) $\left(7x+y-\frac{1}{2}\right)$ d) $-\left(7x+y-\frac{1}{2}\right)$

6 Simplify an expression.

Example 7 Simplify each expression.

a) $7+2(x-1)$ b) $7-2(x-1)$ c) $7-(x-1)$

d) $2+(6a-3b)-(6a+3b)$ e) $2(4a+5)-8(a-1)$ f) $11-(x^2-3)+(6x^2+4)+x$

Answers: 5a) $6x-8$ b) $-3x-12$ c) $-45+5y$ d) $2x-4y$ e) $6a-8b+4$ f) $-\frac{3}{2}x+2y-1$ 6a) $3x-4$ b) $-3x+4$ c) $7x+y-\frac{1}{2}$ d) $-7x-y+\frac{1}{2}$ 7a) $5+2x$ b) $9-2x$ c) $8-x$ d) $2-6b$ e) 18 f) $5x^2+x+18$

Name:
Instructor:

Date:
Section:

Practice Set 2.1

Given $3x - 4 - 8x + 3$, use the choices below to fill in each blank.

| terms | coefficient | constant(s) | variable |
| equation | like terms | combine | expression |

1. $3x - 4 - 8x + 3$ is an example of a(n) _____ .

2. $3x - 4 - 8x + 3$ contains four _____ .

3. The _____ of the third term is -8.

4. The terms -4 and 3 are _____ .

5. If we _____ like terms we get $-5x - 1$.

Combine like terms when possible. If not possible, rewrite the expression as is.

6. $y + y + 1$

7. $-3x - 4x - 5$

8. $\dfrac{1}{2}t + \dfrac{1}{4}t$

9. $9.1 - 0.4x + 1.7x$

10. $4x - \dfrac{3}{8}y + \dfrac{2}{3}y - 1$

11. $-2xy + 2x - xy - x^2$

12. $4x - 7 - x + 4 - 3x + 3$

13. $3x^2 + 3x - x^2 + x$

6. _____
7. _____
8. _____
9. _____
10. _____
11. _____
12. _____
13. _____

Use the distributive property to remove parentheses.

14. $4(8x - 2)$

15. $-\dfrac{1}{3}(2 - 6x)$

16. $-(x + y - 7)$

17. $100(2x - 0.04)$

14. _____
15. _____
16. _____
17. _____

Practice Set 2.1

Simplify.

18. $18 - (8x - 18)$

19. $10\left(\dfrac{2}{5}x - 3 + \dfrac{9}{10}x - \dfrac{1}{2}\right)$

20. $-\left(9x - \dfrac{1}{2}\right) + \dfrac{3}{2}$

21. $12(t - 1) - 3(4t + 5)$

18. _____

19. _____

20. _____

21. _____

Name: Date:
Instructor: Section:

2.2 The Addition Property of Equality

Objectives
1. Identify linear equations.
2. Check solutions to equations.
3. Identify equivalent equations.
4. Use the addition property to solve equations.
5. Solve equations by doing some steps mentally.

Key Vocabulary
equation, linear equations, solution to an equation, equivalent equations, solve an equation, addition property of equality

1 Identify linear equations.

Example 1 Determine whether each is a linear equation.

a) $3x+4$ b) $3x+4=2$ c) $3x^2+4=2$ d) $3x+4=2x-1$

2 Check solutions to equations.

Example 2 Determine whether the given value is a solution to the given equation.

a) $5x-3=7$; 2

b) $6-(3x-2)=-(2x+16)$; 24

c) $4(x-1)=2(2x+3)$; 0

d) $10-6x=-16x$; -1

3 Identify equivalent equations.

Example 3 Match each equation on the left with an equivalent equation on the right.

a) $7x-3=5x+1$ _____   A. $12x=2$

b) $7x-1=1-5x$ _____   B. $x=4$

c) $3x=12$ _____   C. $2x=4$

d) $-x=6$ _____   D. $x=-6$

E. $x=6$

Answers: 1a) no b) yes c) no d) yes 2a) yes b) yes c) no d) yes 3a) C b) A c) B d) D

Examples 2.2

4. Use the addition property to solve equations.

Example 4 Use the addition property to solve each equation.

a) $x + 3 = 24$

b) $4 + x = 1$

c) $y - 4 = 14$

d) $15 = y - 3$

5. Solve equations by doing some steps mentally.

Example 5 Solve each equation by performing some steps mentally.

a) $x - 3 = 24$

b) $3 + x = 14$

c) $y + 14 = 4$

d) $x - 3 = -10$

Answers: 4a) 21 b) −3 c) 18 d) 18   5a) 27 b) 11 c) −10 d) −7

Name:
Instructor:

Date:
Section:

Practice Set 2.2

Match the item on the left with the best match on the right.

1. $3x = 9$ _____   **A.** solution is $x = 0$

2. $9 = x - 3$ _____   **B.** solution is $x = 6$

3. $x - 6$ _____   **C.** equivalent to $x = 12$

4. $x + 6 = 6$ _____   **D.** equivalent to $x = 3$

5. $3 + x = 9$ _____   **E.** not a linear equation

6. $x - 6 = 6$ _____

Solve each equation and check your solution.

7. $y + 4 = 14$ 8. $17 + r = -10$ 7. _____

                                                                                                                                 8. _____

9. $-11 = x - 11$ 10. $21 = x + 1$ 9. _____

                                                                                                                                 10. _____

11. $x + 16 = 16$ 12. $y - 4 = 14$ 11. _____

                                                                                                                                 12. _____

13. $x - 1 = 2$ 14. $12 = s - 20$ 13. _____

                                                                                                                                 14. _____

15. $-24 = x - 4$ 16. $18.2 + x = -17$ 15. _____

                                                                                                                                 16. _____

17. $0.29 = x + 0.29$ 18. $-0.7 - c = 1$ 17. _____

                                                                                                                                 18. _____

Challenge

19.   Solve for $\triangle$: $\triangle - 7 = 5$ 19. _____

20.   Solve for ♥: $-8.4 = -0.3 -$ ♥   20. _____

Name:
Instructor:

Date:
Section:

2.3 The Multiplication Property of Equality

> **Objectives**
> 1. Identify reciprocals.
> 2. Use the multiplication property to solve equations.
> 3. Solve equations of the form $-x = a$.
> 4. Do some steps mentally when solving equations.
>
> **Key Vocabulary**
> reciprocals, multiplication property of equality

1 Identify reciprocals.

Example 1 Determine the reciprocal of each number.

    a) $\dfrac{2}{3}$ b) 3 c) $-\dfrac{1}{3}$ d) -5

2 Use the multiplication property to solve equations.

Example 2 Determine the coefficient of each term.

    a) $3x$ b) $4x$ c) $-6x$ d) $\dfrac{x}{3}$

Example 3 Use the multiplication property to solve each equation.

    a) $3x = 24$ b) $4x = 1$ c) $-6x = -24$ d) $8 = \dfrac{x}{3}$

3 Solve equations of the form $-x = a$.

Example 4 Solve each equation.

    a) $-x = 10$ b) $-x = -10$ c) $12 = -x$ d) $-\dfrac{1}{2} = -x$

4 Do some steps mentally when solving equations.

Example 5 Solve each equation by performing some steps mentally.

    a) $11x = 55$ b) $-x = \dfrac{2}{3}$ c) $-7x = 56$ d) $-8 = \dfrac{-x}{4}$

___

Answers: 1a) $\dfrac{3}{2}$ b) $\dfrac{1}{3}$ c) -3 d) $-\dfrac{1}{5}$ 2a) 3 b) 4 c) -6 d) $\dfrac{1}{3}$ 3a) 8 b) $\dfrac{1}{4}$ c) 4 d) 24   4a) -10 b) 10 c) -12 d) $\dfrac{1}{2}$
5a) 5 b) $x = -\dfrac{2}{3}$ c) -8 d) 32

Name:
Instructor:

Date:
Section:

Practice Set 2.3

Solve each equation and check your solution.

1. $2y = 14$
2. $-13 = 39r$
3. $-x = -\dfrac{3}{8}$
4. $4x = -14$
5. $-0.15y = 2.25$
6. $20s = -12$
7. $21x = 1$
8. $1.6 = -1.92n$
9. $20 = -\dfrac{t}{8}$
10. $100w = 0.9$
11. $4s = -\dfrac{12}{21}$
12. $-r = -22$
13. $-\dfrac{c}{5} = 15$
14. $\dfrac{7}{8}y = -14$
15. $-3x = \dfrac{9}{11}$
16. $-10.2 = 0.3z$
17. $-\dfrac{2}{3}x = 0$
18. $-\dfrac{4}{5}y = -\dfrac{36}{25}$

1. _____
2. _____
3. _____
4. _____
5. _____
6. _____
7. _____
8. _____
9. _____
10. _____
11. _____
12. _____
13. _____
14. _____
15. _____
16. _____
17. _____
18. _____

Challenge

19. Solve for ♥: $\dfrac{11}{20}♥ = -\dfrac{33}{10}$

20. Solve for △: $-11 = \dfrac{-△}{7}$

19. _____
20. _____

Name:
Instructor:
Date:
Section:

2.4 Solving Linear Equations with a Variable on Only One Side of the Equation

Objectives
1. Solve linear equations with a variable on only one side of the equation.
2. Solve equations containing decimal numbers or fractions.

1 Solve linear equations with a variable on only one side of the equation.

Example 1 Solve each equation.

a) $13 - x = 7$ b) $32x + 10 = -10$ c) $12.6 = 4.6x - 18.68$ d) $\dfrac{2}{5}(x - 10) = 16$

2 Solve equations containing decimal numbers or fractions.

Review Decimals

Example 2 Find each product mentally.

a) $(2.7)(10)$ b) $100(0.92x)$ c) $10(3x - 0.4)$ d) $100(3.8 + 0.42x)$

Review LCD

Example 3 Find the LCD of the following fractions.

a) $\dfrac{2}{9}, \dfrac{3}{5}$ b) $\dfrac{2}{3}, \dfrac{4}{3}, \dfrac{3}{4}$

Example 4 Solve each equation by first clearing all decimals or fractions.

a) $\dfrac{2}{3}x - \dfrac{4}{3}x = \dfrac{3}{4}$ b) $0.32x + 0.10 = -0.10$

c) $8.5 = -2.7 + 3.2x$ d) $\dfrac{1}{3}(x + 4) = 7$

Answers: 1a) 6   b) $-\dfrac{5}{8}$ c) 6.8   d) 50   2a) 27   b) 92x   c) 30x − 4   d) 380 + 42x   3a) 45   b) 12   4a) $-\dfrac{9}{8}$ b) −0.625   c) 3.5   d) 17

Name:
Instructor:

Date:
Section:

Practice Set 2.4

Match each equation on the left with an equivalent equation on the right.

1. $\dfrac{3}{4}x - \dfrac{x}{4} - 3 = 2$ _____ A. $3x - 12 = 8$

2. $\dfrac{4}{3} = \dfrac{x-4}{2}$ _____ B. $3x - 10x - 120 = 8$

3. $0.3x - 0.8 = 1.2$ _____ C. $5x - 3 = 4$

4. $\dfrac{1}{2}(x-6) = 4$ _____ D. $3x - x - 12 = 8$

5. $0.03x - 0.1(x+12) = 0.08$ _____ E. $2x - 12 = 8$

6. $0.05x - 0.3 = 0.4$ _____ F. $x - 6 = 8$

Solve each equation.

7. $9x - 4 = 14$

8. $7 - x = 10$

7. _____

8. _____

9. $-8x + 3 = 27$

10. $13 = 9 - c$

9. _____

10. _____

11. $-\dfrac{11}{18} - \dfrac{x}{9} = \dfrac{1}{6}$

12. $50 = 4 - 8t + 2$

11. _____

12. _____

13. $9(x+1) = 12$

14. $-\dfrac{2}{9}(y-7) = \dfrac{2}{3}$

13. _____

14. _____

15. $1.95x + 0.05x = -8$

16. $\dfrac{x}{12} - \dfrac{5x+1}{2} = \dfrac{1}{12}$

15. _____

16. _____

17. $1.2x + x = 0.28 - 0.10$

18. $0.05x - 0.3 = 0.2$

17. _____

18. _____

Name:
Instructor:

Date:
Section:

2.5 Solving Linear Equations with the Variable on Both Sides of the Equation

Objectives
1. Solve equations with the variable on both sides of the equations.
2. Solve equations containing decimal numbers or fractions.
3. Identify identities and contradictions.

Key Vocabulary
conditional equation, identity, contradiction

1 Solve equations with the variable on both sides of the equation.

Example 1 Solve each equation.

a) $4x + 6 = 50 + 3x$

b) $3x - 7 = 5x - 5$

c) $6(y + 2) = 2(y - 7)$

d) $2(3 - r) = 3(3 + 2r) + 8r$

2 Solve equations containing decimal numbers or fractions.

Example 2 Solve each equation.

a) $\dfrac{9}{2} - \dfrac{m}{10} = \dfrac{m}{20}$

b) $4.8c = 12 - 2.4(c - 4)$

c) $3.4(2x + 1.4) = 10.2(x - 4.2)$

d) $\dfrac{9}{8}(x - 4) = \dfrac{3(2x - 1)}{4}$

3 Identify identities and contradictions.

Example 3 Solve each equation. Identify any identities and contradictions.

a) $4x + 4 = 5x - (x - 4)$

b) $1.5(x + 2) = 0.3(5x - 10)$

c) $8 - 2y = \dfrac{1}{2}(8 - 4y)$

d) $6x - 3 = 0.2(30x - 15)$

Answers: 1a) 44 b) -1 c) $-\dfrac{13}{2}$ d) $-\dfrac{3}{16}$ 2a) 30 b) 3 c) 14 d) -10 3a) all real numbers; identity b) no solution; contradiction c) no solution; contradiction d) all real numbers; identity

Name:
Instructor:

Date:
Section:

Practice Set 2.5

Match the equation on the left to its best description on the right.

1. $4x - 10 = 14 - 4x$ _____
2. $2(2x - 5) = 4x - 10$ _____
3. $2(7 - 2x) = 14 - 4x$ _____
4. $2(5 - 2x) = 2(7 - 2x)$ _____

A. identity

B. contradiction

C. conditional equation

Solve each equation. Identify any identities and contradictions.

5. $11y - 6 = 8y + 3$

6. $9x + 2 = 12 + 3x$

5. _____

6. _____

7. $2(2x - 3) = 22 - 3x$

8. $15 - 3c = 3(5 - c)$

7. _____

8. _____

9. $0.9 - 0.45p = 0.85 + 0.05p$

10. $18(r - 2) = 4(r - 9)$

9. _____

10. _____

11. $\dfrac{x + 3}{6} = \dfrac{4x - 1}{2}$

12. $3(3 + 2y) - 8y = 2(3 - y)$

11. _____

12. _____

13. $0.81 - 0.06t = 0.18t + 0.03(4t - 9)$

14. $0.625 - 0.5z = 0.25z$

13. _____

14. _____

Practice Set 2.5

15. $2x + \dfrac{2}{3} = \dfrac{6}{5}x - 4$

16. $\dfrac{x+5}{4} - \dfrac{x-7}{6} = \dfrac{7x-1}{12}$

15. _____

16. _____

17. $\dfrac{4}{3}(2x+1) = \dfrac{6}{5}(x+2) - \dfrac{1}{3}$

18. $0.8(2x+1.4) = 1.2(2x-8.4)$

17. _____

18. _____

Challenge

19. Solve $\dfrac{\Omega-5}{24} - \dfrac{\Omega+7}{8} = \dfrac{\Omega-1}{12}$ for Ω.

19. _____

20. Solve $9(\Delta-2) + \Delta = 3(\Delta-7)$ for Δ.

20. _____

Name:
Instructor:

Date:
Section:

2.6 Formulas

> **Objectives**
> 1. Use the simple interest formula and the distance formula.
> 2. Use geometric formulas.
> 3. Solve for a variable in a formula.
>
> **Key Vocabulary**
> formula, evaluate a formula, simple interest formula, distance formula, perimeter, area, quadrilateral, circumference, radius, diameter, pi (π), volume

1 Use the simple interest formula and the distance formula.

Example 1 Mindy Lou borrowed $1200 from her mother, who would charge 4.5% simple annual interest for the 2-year loan.
a) How much interest will Mindy pay her mother?
b) How much in all will Mindy pay her mother for the loan?

Example 2 Joe Landis traveled for 6 h 15 min from Chicago, Illinois to Augusta, Missouri at an average speed of 68 miles per hour. Determine the distance Joe traveled.

2 Use geometric formulas.

Example 3 Nita Moellering is preparing for Halloween. She wants to hang jack-o-lantern lights around the edge of her rectangular deck that measures 5 ft by 8 ft. How many feet of lights does she need?

Example 4 Refer to the information given in Example 3 above. If Nita had a circular deck with a diameter of 8 ft, how many feet of lights would she need?

3 Solve for a variable in a formula.

Example 5 Solve for the indicated variable.

a) $i = prt$, for r

b) $PV = NRT$, for V

c) $3x - y = 12$, for y

d) $P = a + b + c$, for a

Answers: 1a) $108 b) $1308 2. 425 mi 3. 26 ft 4. approximately 25.13 ft, or 25 ft 1.6 in. 5a) $r = \dfrac{i}{pt}$ b) $V = \dfrac{NRT}{P}$
c) $y = 3x - 12$ d) $a = P - b - c$

Name:
Instructor:

Date:
Section:

Practice Set 2.6

Use the choices below to fill in each blank.

    perimeter   trapezoid   circumference   triangle

    radius   area   quadrilateral(s)   diameter

1. A square, rectangle, and _____ are types of _____.

2. If the _____ of a circle is 8 in., then the circle's _____ is 16 in.

3. The _____ is the distance around a quadrilateral, whereas the _____ is the distance around a circle.

4. The formula for the _____ of a rectangle is $A = LW$.

Use the formula to find the value of the variable indicated. Use a calculator to save time and where necessary, round your answer to the nearest hundredth.

5. $P = 2L + 2W$; find P when $L = 10$ and $W = 0.8$

6. $m = a + b + c$; find m when $a = 43.2$, $b = 90$, and $c = 46.8$

5. _____

6. _____

7. $A = \dfrac{1}{2}(B + b)h$; find A when $B = 1.2$, $b = 10$, and $h = 14$

8. $S = 4\pi r + 2\pi r^2$; find S when $r = 10$

7. _____

8. _____

Solve for the indicated variable.

9. $2x - 3y = 12$, for y

10. $P = 2(L + W)$, for W

9. _____

10. _____

11. $V = \dfrac{1}{3}\pi r^2 h$, for h

12. $ax + by = c$, for y

11. _____

12. _____

Practice Set 2.6

Problem Solving

Use the appropriate formula to solve each problem.

13. Rhonda Sanchez put $350 into a savings account paying $3\frac{3}{4}$% simple interest per year. How much interest will she earn in 4 years?

    13. _____

14. Jacob Stauffer rode his bike for 3 hours on the Katy Trail at an average speed of 8 miles per hour. How far did he travel?

    14. _____

15. A bed & breakfast in Oshkosh, Wisconsin, just had a rectangular sign delivered that is to hang directly above the front door. If the sign is 4 ft long and $2\frac{3}{4}$ ft high, how much area above the door will the sign cover?

    15. _____

16. Ms. Stigall, an 8th grade social studies teacher, has a bulletin board in her classroom that is 3.5 ft by 3.5 ft square. If she wants to hang a decorative border around the perimeter of the bulletin board, how long is the border?

    16. _____

17. Refer to Problem 16 above. Ms. Stigall also wants to cover the entire bulletin board with decorative paper. How many square feet of paper does she need?

    17. _____

18. A speaker to a DVD sound system is in the shape of a rectangular solid. It measures 15 in. wide, 10 in. deep, and 1.5 ft tall. Find the volume of the speaker in cubic inches.

    18. _____

19. Charles Schowalter took out a $2000 loan from his friend at $4\frac{1}{4}$% simple interest for $3\frac{1}{2}$ years.

    a) How much will Charles pay in interest?
    b) How much will Charles end up paying back his friend?

    19. a) _____

    b) _____

Challenge

20. Pam entered a triathlon in which she ran for 80 min at an average speed of $4\frac{1}{2}$ mph, biked 10 mi at an average speed of 35 mph, and swam 3 mi at an average speed of 2 mph. How many miles in all did Pam run, bike, and swim?

    20. _____

Name:
Instructor:
Date:
Section:

2.7 Ratios and Proportions

Objectives
1. Understand ratios.
2. Solve proportions using cross-multiplication.
3. Solve applications.
4. Use proportions to change units.
5. Use proportions to solve problems involving similar figures.

Key Vocabulary
ratio, terms of the ratio, proportion, extremes, means, cross-multiplication, similar figures

1 Understand ratios.

Example 1 Flour ratios A European bakery's recipe for multigrain bread calls for 20 cups of wheat flour, 60 cups of all-purpose white flour, and 15 cups of rye flour. Determine the indicated ratio and write the ratio in lowest terms.

a) Find the ratio of number of cups of wheat flour to the number of cups of all-purpose white flour.

b) Find the ratio of number of cups of rye flour to the total amount of flour in the recipe.

Example 2 Determine the following ratios. Write each ratio in lowest terms.

    a)   3 ft to 6 ft   **b)**   10 cups to 4 cups

    c)   2 h to 40 min   **d)**   3 ft to 2 yd

Example 3 Write each ratio in Example 2 as some quantity to 1.

2 Solve proportions using cross-multiplication.

Example 4 Solve the following proportions by cross-multiplying.

    a) $\dfrac{x}{4} = \dfrac{18}{24}$ **b)** $\dfrac{9}{48} = \dfrac{c}{3}$

    c) $\dfrac{-12}{y} = \dfrac{4}{9}$ **d)** $\dfrac{x}{7.5} = \dfrac{2}{2.5}$

_____

Answers: 1a) 1 : 3 b) 3 : 19 2a) 1 : 2 b) 5 : 2 c) 3 : 1 d) 1 : 2 3a) 0.5 : 1 b) 2.5 : 1 c) 3 : 1 d) 0.5 : 1 4a) 3 b) $\dfrac{9}{16}$ c) -27 d) 6

Examples 2.7

3 Solve applications.

Example 5 The legend on a map says that 0.5 inch represents 4 miles. If the distance between two towns on the map is 6.75 inches, what is the distance, in miles, between the two towns?

Example 6 Jason usually uses 8 scoops of coffee grinds for his 10-cup coffee maker. If he only needed to make 6 cups of coffee one morning, approximately how many scoops of coffee grinds does he need?

4 Use proportions to change units.

Example 7 Before Marleen Dreher traveled to Germany in 2013, the bank teller told her that $1 U.S. could be exchanged for 1.30 euros.

a) How many euros can she get if she exchanged $500 U.S.?

b) Marleen paid 85 euros for a silk scarf in the historic city of Alsfeld. Using the exchange rate given, determine the cost of the scarf in U.S. dollars.

5 Use proportions to solve problems involving similar figures.

Example 8 The figures below are similar. Find the length of the side indicated by x.

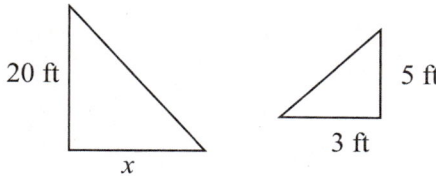

Answers: 5. 54 mi 6. almost 5 scoops 7a) 650 euros b) $65.38 8. 12 ft

Name:
Instructor:

Date:
Section:

Practice Set 2.7

Match the proportion on the left to its best match on the right. Note: Some proportions may have more than one match.

1. $\dfrac{x}{6} = \dfrac{3}{12}$ _____

2. $\dfrac{6}{12} = \dfrac{x}{3}$ _____

3. $\dfrac{3}{x} = \dfrac{6}{12}$ _____

4. $\dfrac{3}{6} = \dfrac{x}{12}$ _____

A. product of means is $12x$

B. product of means is 36

C. product of extremes is $12x$

D. product of extremes is 18

E. product of extremes is 36

Determine the following ratio. Write the ratio in lowest terms.

5. 4 hours to 10 hours

6. 12 females to 9 males

7. 6 teaspoons to 4 tablespoons
(Hint: 1 tablespoon = 3 teaspoons)

8. 5 hours to 30 minutes

5. _____

6. _____

7. _____

8. _____

Write the indicated ratio as some quantity to 1.

9. 4 hours to 10 hours

10. 62.8 in. to 20 in.

9. _____

10. _____

Solve each proportion for the variable by cross-multiplying.

11. $\dfrac{x}{4} = \dfrac{5}{32}$

12. $\dfrac{2.5}{1.8} = \dfrac{y}{-0.3}$

11. _____

12. _____

Problem Solving

13. The legend on a map says that 0.25 inch represents 5 miles. If the distance between two towns on the map is 8.75 inches, what is the distance, in miles, between the two towns?

13. _____

Practice Set 2.7

14. A recipe that makes two loaves of sourdough bread calls for 7 cups of flour. How many loaves of bread will 10.5 cups of flour yield?

14. _____

15. Given that there are 2.54 cm in one inch, what is the height, in centimeters, of a young woman who is 64 inches tall?

15. _____

16. A group of college students backpacked through Europe on their senior trip. At that time, the $1 U.S. could be exchanged for 1.40 euros.
a) How many euros can a student get if he exchanged $350 U.S.?
b) Marcelin paid 45 euros for a book in Prague. Using the exchange rate given, determine the cost of the book in U.S. dollars.

16. (a) _____

(b) _____

17. The label of a concentrated household cleaner says to use $\frac{1}{4}$ cup of the liquid for every 32 fl oz of water. How much cleaner is needed for 80 fl oz of water?

17. _____

18. The figures below are similar. Find the length of the side indicated by x.

18. _____

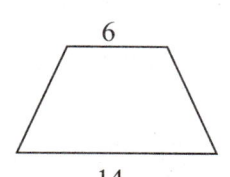
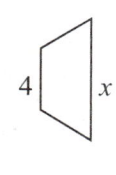

Copyright © 2015 by Pearson Education, Inc.   Angel/Runde, *Elementary Algebra for College Students*, 9e   55

Name:
Instructor:

Date:
Section:

2.8 Inequalities in One Variable

Objectives
1. Solve linear inequalities and graph the solution on a number line.
2. Write inequality solutions in interval notation.
3. Solve linear inequalities that have all real numbers as their solution, or have no solution.

Key Vocabulary
inequality, sense (order) of an inequality

1 Solve linear inequalities and graph the solution on a number line.

Example 1 Solve the inequality and graph the solution on a number line.

a) $x + 8 < 5$

b) $-10t < 6(t - 2)$

c) $\dfrac{y}{4} < \dfrac{y}{6} + \dfrac{5}{2}$

2 Write inequality solutions in interval notation.

Example 2 Solve the inequality. Graph the solution on a number line and represent the solution in interval notation.

a) $3r - 6 > 9$

b) $7.0x - 7.6 \leq 2.4x + 6.2$

c) $12(3 - r) > r + 6$

3 Solve linear inequalities that have all real numbers as their solution, or have no solution.

Example 3 Solve the inequality Graph the solution on a number line and represent the solution in interval notation.

a) $x + 5 < x - 8$

b) $4(p + 2) > 2(2p - 5)$

c) $\dfrac{4}{5} - y > \dfrac{1}{10} - y$

d) $1.2(3 - z) \geq 2(7 - 0.6z)$

Answers:
1. a) $x < -3$ b) $t > \dfrac{3}{4}$ c) $y < 30$ 2. a) $r > 5$, $(5, \infty)$ b) $x \leq 3$, $(-\infty, 3]$ c) $r < \dfrac{30}{13}$, $\left(-\infty, \dfrac{30}{13}\right)$ 3. a) no solution b) $(-\infty, \infty)$ c) $(-\infty, \infty)$ d) no solution

56 Angel/Runde, *Elementary Algebra for College Students*, 9e

Practice Set 2.8

Match the inequality on the left to its solution on the right.

1. $-x < 5$ _____
2. $x + 5 > x - 5$ _____
3. $-x < -5$ _____
4. $-x + 5 < -(x - 5)$ _____
5. $-x > -5$ _____

A. (number line: open circle at 5, arrow left; 3 4 5 6 7)
B. (number line: open circle at 5, arrow right; 3 4 5 6 7)
C. (number line: open circle at -5, arrow right; -7 -6 -5 -4 -3)
D. (number line: shaded entire line; -2 -1 0 1 2)
E. (number line: no shading; -2 -1 0 1 2)

Solve each inequality. Graph the solution on a number line and represent the solution in interval notation.

6. $y + 7 > 9$
7. $20 \leq x + 12$
8. $-5 < -3 - x$
9. $-y > -4 - y$
10. $7x - 12 \geq -5$
11. $1 - 5x < 10x - 11$
12. $-6(x - 1) > 2(3 - 3x)$
13. $12(2m + 3) \leq 6(3m - 1)$
14. $4.6 - 13x > 10.6x - 13.4$
15. $\dfrac{n}{12} \leq \dfrac{n}{18} + \dfrac{5}{6}$
16. $6.2 + 2.4r \geq 7r - 7.6$
17. $\dfrac{x+5}{4} - \dfrac{x-7}{6} < \dfrac{7x-1}{12}$

6. _____
7. _____
8. _____
9. _____
10. _____
11. _____
12. _____
13. _____
14. _____
15. _____
16. _____
17. _____

Notes:

Chapter 2 Vocabulary Reference Sheet

Term	Definition	Example
term	A constant, variable, or a product of a constant and variable or variables	In $2x - 3y + 5$, $2x$, $-3y$, and 5 are terms.
(numerical) coefficient	The numerical factor of a term	In $-3y$, the coefficient is -3.
constant (term)	A real number standing alone	In $2x - 3y + 5$, the constant is 5.
like terms	Terms that have the same variables with the same exponents, respectively.	In $2x - 4 - 3y - 5x + 1$, there are 2 pairs of like terms: $2x$ and $-5x$; -4 and 1.
combine like terms	To add or subtract the coefficients of the like terms	Combine like terms: $$2x - 4 - 3y - 5x + 1$$ $$2x - 5x - 4 + 1 - 3y$$ $$-3x - 3 - 3y$$
(algebraic) expression	A term or combination of terms being added, subtracted, multiplied, or divided	$2x - 4$ is an expression, but $2x - 4 = 5$ is not an expression.
equation	A statement in which two expressions are set equal to each other	$2x - 4 = 5$ or $2x^2 - 4x = 0$
linear equation	An equation of the form $ax + by = c$, where $a \neq 0$	$2x - 4 = 5$ is linear, but $2x^2 - 4x = 0$ is not linear.
solution to an equation	The number or numbers that when substituted for the variable or variables make an equation a true statement	$x = \dfrac{9}{2}$ is a solution to $2x - 4 = 5$ since $$2\left(\dfrac{9}{2}\right) - 4 = 5$$ $$9 - 4 = 5$$ $$5 = 5$$ is a true statement.
equivalent equations	Two or more equations that have the same solution	$2x - 4 = 5$ is equivalent to $2x = 9$.
reciprocal	The multiplicative inverse of a real number	Example: $\dfrac{2}{3}$ and $\dfrac{3}{2}$; $\dfrac{2}{3} \cdot \dfrac{3}{2} = 1$
conditional equation	An equation that has a single value for a solution	$x = \dfrac{9}{2}$ is a conditional equation since it only has the solution $x = \dfrac{9}{2}$.
identity	An equation that is true for infinitely many values of the variable	$2x - 4 = 2(x - 2)$ is an identity since it has infinitely many solutions. Notice: $$2x - 4 = 2x - 4$$ $$2x - 2x - 4 = 2x - 2x - 4$$ $$-4 = -4$$ and the result is a true statement.
contradiction	An equation that is not true for any value of the variable.	$2x - 4 = 2(x - 3)$ is a contradiction since it has no solution. Notice: $$2x - 4 = 2x - 6$$ $$2x - 2x - 4 = 2x - 2x - 6$$ $$-4 = -6$$ and the result is a false statement.

Chapter 2 Vocabulary

formula	An equation commonly used to express a specific relationship mathematically	$P = 2L + 2W$, $A = LW$, and $A = \pi r^2$ are all formulas.
simple interest formula	interest = principal · rate · time, or $i = prt$	$i = prt$ is the formula used to solve problems involving simple interest. i is interest, p is principal, r is the interest rate, and t is the number of years.
perimeter	The sum of the lengths of the side of a figure	A rectangle that is 4 ft by 3 ft has a perimeter of $P = 2L + 2W$, so $P = 2(4) + 2(3) = 14$ ft.
area	The total surface area within a figure's boundaries	A rectangle that is 4 ft by 3 ft has an area of $A = LW$. So $A = (4)(3) = 12$ ft².
quadrilateral	A 4-sided figure	A rectangle, square, trapezoid, rhombus, and parallelogram are types of quadrilaterals.
circumference	The length (or perimeter) of the curve that forms the circle; $C = 2\pi r$ or $C = \pi d$	A circle with radius $r = 4$ has a circumference of $C = 2\pi(4) = 8\pi \approx 25.12$.
radius	A line segment from the center of a circle to any point on the circle	In the circle below, the dashed line segment is a diameter and the solid line segment is a radius.
diameter	A line segment through the center of a circle whose endpoints both lie on the circle; the length of the diameter is twice the length of the radius	
ratio	A quotient of two quantities	4:5 is also $\frac{4}{5}$.
terms of the ratio	The two quantities that are being divided in a ratio	In $\frac{12}{20}$, 12 and 20 are the terms.
proportion	An equation in which two ratios are set equal to each other	$\frac{x}{5} = \frac{12}{20}$
extremes	Cross terms of a ratio, specifically a and d in $\frac{a}{b} = \frac{c}{d}$	In $\frac{x}{5} = \frac{12}{20}$, x and 20 are the extremes.
means	Cross terms of a ratio, specifically b and c in $\frac{a}{b} = \frac{c}{d}$	In $\frac{x}{5} = \frac{12}{20}$, 5 and 12 are the means.
cross-multiplication	Product of the means, bc, equals product of the extremes, ad	To solve $\frac{x}{5} = \frac{12}{20}$, we can cross-multiply to get $20x = 60$, then $x = 3$.
similar figures	Figures in which corresponding angles are equal in measure and corresponding sides are proportional	The following are similar triangles:
inequality	A mathematical statement containing one or more of the following: $<, >, \leq, \geq$	$2x - 8 \geq 6$, $5 > 2$, $x^2 + 2 \geq 6$
interval notation	A method for representing solutions to inequalities using parentheses, brackets, or a combination of the two	In interval notation, $x > 3$ is represented as $(3, \infty)$ and $x \leq 5$ is represented as $(-\infty, 5]$.

Name:
Instructor:

Date:
Section:

Chapter 2 Practice Test A

Simplify.

1. $-3(5x-4)$

    1._____

2. $4x - y - 7x + 4 - 10x$

    2._____

3. $\dfrac{4}{5}x + \dfrac{7}{8} - \dfrac{9}{10}x - \dfrac{7}{12}$

    3._____

4. $9x - (4x^2 - 1) + 8x - 7$

    4._____

Solve.

5. $5(x+4) = 10$

    5._____

6. $-(r-5) + 3r = 2(r+1) + 3$

    6._____

7. $7.3(2x-1) + 4.9 = -1.4(x+3)$

    7._____

8. $\dfrac{3}{4}y - \dfrac{1}{9} = \dfrac{7}{9}y - 4$

    8._____

9. Solve for y.

    $6x - y = 3$

    9._____

10. Solve for h.

    $A = \dfrac{1}{2}(B+b)h$

    10._____

Solve the inequality. Graph the solution on a number line and represent the solution in interval notation.

11. $3(5x+6) > 11x - 8$

    11._____

12. $\dfrac{2}{5}x - \dfrac{5}{6} \leq \dfrac{1}{3} + x$

    12._____

Practice Test 2A

13. The figures below are similar. Find the length of the side indicated by *s*.

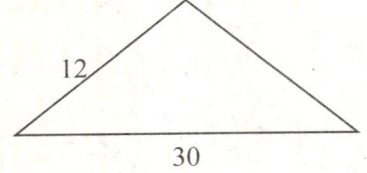
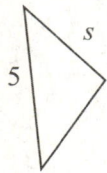

13. _____

14. "Sour salt," or citric acid, is used to enhance the flavor of bread. A recipe for rye bread calls for $\frac{1}{4}$ teaspoon sour salt for every $2\frac{1}{4}$ cups rye flour. How much sour salt is needed if 9 cups of rye flour is used?

14. _____

15. Irma wants to send her grandson, Max, a birthday gift. She was told that the discount rate would apply if the box's volume was no larger 1200 in.³ If the box containing Max's birthday gift is 25 in. long, 8 in. wide, and 4 in. deep, what is the box's volume? Can she send the box at the discount rate?

15. _____
_____

16. The Cheesecake Company made a pumpkin cheesecake in a round 12-in. pan (12-in. diameter). If pecans 1 inch in length are to be placed around the circumference of the cheesecake, how many pecans will fit around the cake? Round to the nearest pecan.

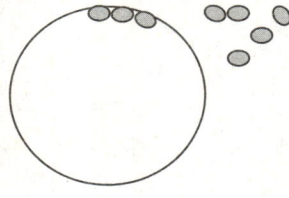

16. _____

17. The last time Jade visited his friend who was away at college, it took him $3\frac{1}{2}$ hours to travel 217 miles. At this speed, how long will it take him to get to his Aunt Robin's house if she lives 50 miles away? Round to the nearest minute.

17. _____

18. Terra Jones borrowed $550 from her friend Mary Margaret. Terra was charged an interest rate of $3\frac{1}{2}$ % for this 1-year loan. How much in all will Terra pay Mary Margaret after 1 year?

18. _____

Chapter 2 Practice Test B

1. Use the distributive property to simplify $-2(5x-1)$.
   a) $-10x-1$ b) $-10x+2$ c) $-7x+2$ d) $-10x+2$

2. Simplify $4a-b+3a-5b$.
   a) $7a-6b$ b) $7a+6b$ c) $7a^2-6b^2$ d) $7a^2+6b^2$

3. Simplify.
   $$\frac{3}{8}-\frac{1}{4}x-\frac{3}{4}+\frac{7}{8}x$$
   a) $5x-3$ b) $3-5x$ c) $\frac{3}{8}-\frac{5}{8}x$ d) $\frac{5}{8}x-\frac{3}{8}$

4. Solve $24x-6=4x+4$.
   a) -2 b) $-\frac{1}{2}$ c) $\frac{1}{2}$ d) 2

5. Solve $1.2(3x-2)=6.4x+0.4$.
   a) -1 b) $-\frac{6}{7}$ c) $\frac{6}{7}$ d) 1

6. Solve $-(x-3)=5-x$.
   a) 4 b) -4 c) all real numbers   d) no solution

7. Solve.
   $$\frac{2}{5}x-\frac{3}{10}=\frac{2}{3}-\frac{1}{10}x$$
   a) $\frac{15}{29}$ b) $\frac{9}{11}$ c) $\frac{11}{9}$ d) $\frac{29}{15}$

8. Solve.
   $$\frac{5}{7}=\frac{x}{-35}$$
   a) -25 b) -1 c) 1 d) 25

9. Solve $5x-3y=12$ for y.
   a) $y=\frac{5}{3}x+4$ b) $y=\frac{5}{3}x-4$ c) $y=15-5x$ d) $y=5x-15$

10. Solve for b.
    $$A=\frac{1}{2}(B+b)h$$
    a) $b=\frac{2A}{B}-h$ b) $b=\frac{2A-B}{h}$ c) $b=\frac{1}{2}h-B$ d) $b=\frac{2A}{h}-B$

11. Solve $2x+7>9x-7$ and graph the solution on a number line.
    a)    b)   c)   d)

Practice Test 2B

12. Solve $2(3x+8)-6 > 3(2x-1)+13$ and graph the solution on a number line.

   a) ![number line with open circle at 1]

   b) ![number line with open circle at -1]

   c) ![number line from -2 to 2]

   all real numbers

   d) ![number line from -2 to 2]

   no solution

13. Choose the equation that is equivalent to the proportion $\dfrac{12}{25} = \dfrac{x}{4}$.

   a) $25x = 48$
   b) $12x = 100$
   c) $4x = 300$
   d) $\dfrac{3}{25} = \dfrac{x}{1}$

14. The figures below are similar. Find the length of side x.

   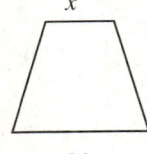
   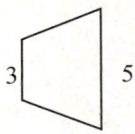

   a) 5
   b) 18
   c) 20
   d) 25

15. A medium pizza at an Italian pizzeria has a diameter of 13 in. Determine the area of the pizza to the nearest square inch.

   a) 41 in.²
   b) 82 in.²
   c) 133 in.²
   d) 531 in.²

16. A box containing a new textbook was delivered to the office of Professor Daniels. The box was 9 in. wide, 12 in. long, and had a 3-in. depth. Find the volume of the box.

   a) 32.4 in.³
   b) 324 in.³
   c) 3240 in.³
   d) 34,000 in.³

17. Cindy Syberg let her sister borrow $550 for a period of 2 years at $2\dfrac{1}{2}$% simple annual interest. What is the total amount Cindy's sister will pay her at the end of the 2-year period?

   a) $27.50
   b) $2750
   c) $577.50
   d) $3300

18. The distance between Jim's house and the university he attends is 85 miles. If Jim's speed was constant and it took him $1\dfrac{1}{4}$ hours to get to his school from his house, what was his speed?

   a) 61 mph
   b) 68 mph
   c) 72 mph
   d) 75 mph

Name:
Instructor:

Date:
Section:

3.1 Changing Application Problems into Equations

Objectives
1. Translate phrases into mathematical expressions.
2. Express the relationship between two related quantities.
3. Write expressions involving multiplication.
4. Translate applications into equations.

Key Vocabulary
consecutive integers, consecutive even integers, consecutive odd integers

1 Translate phrases into mathematical expressions.

Example 1 Express each statement as an algebraic expression.

a) The sum of a number and 6

b) Two less than the height

c) Half the difference between a number and 4

d) Five less than the square of a number

2 Express the relationship between two related quantities.

Example 2 For each part, determine what to let $x = $.

a) The roses cost $20 more than the lilies.

b) Callie is 3 years younger than her sister.

c) The measure of the first angle is twice the second angle.

d) Cindy bought $3\frac{1}{2}$ lb of granola; some was blueberry flavored and the rest was plain.

Answers: 1a) $x + 6$ b) $h - 2$ c) $\frac{1}{2}(x - 4)$ d) $x^2 - 5$ 2a) $x = $ cost of lilies b) $x = $ Callie's sister's age c) $x = $ measure of the second angle d) Let $x = $ number of pounds of plain granola or let $x = $ number of pounds of blueberry granola.

Examples 3.1

3 **Write expressions involving multiplication.**

Example 3 Write each phrase as an algebraic expression.

    a)   The number of feet in m yards

    b)   A $7\frac{1}{2}$% tax on s dollars worth of groceries

    c)   The cost of x CDs at \$14.99 each

    d)   Terra's annual interest rate, r, on her credit card doubled.

4 **Translate applications into equations.**

Example 4 Write each statement as an equation.

    a)   One number is 3 less than twice another number.

    b)   The sum of two consecutive page numbers is 591.

    c)   A New York cab driver's earnings is 30 times that of a Bangladesh rickshaw driver's earnings. If their daily earnings combined is \$92, how much does each driver make in one day?

    d)   The price of the meal with a $7\frac{1}{4}$% tax was \$34.90. Find the price of the meal before tax.

Answers: 3a) $3m$ b) $0.075s$ c) $14.99x$ d) $2r$ 4a) $m = 2n - 3$ b) $x + (x + 1) = 591$ c) $b + 30b = 92$ d) $p + 0.0725p = 34.90$

Name:
Instructor:
Date:
Section:

Practice Set 3.1

Match each phrase on the left to its mathematical translation on the right.

1. Five more than twice a number   _____   A. $5 - 2n$

2. Twice the sum of a number and 5   _____   B. $2n + 5$

3. Five less than twice a number   _____   C. $n^2 - 5$

4. The sum of 5 and a number's square   _____   D. $n^2 + 5$

   E. $2(n + 5)$

   F. $2n - 5$

Express each statement as an algebraic expression.

5. Twice the original price, p

6. Three less than the height, h

7. Half of the weight, w, of the box

8. Two more than the width, w

9. Two-thirds the cost, c, decreased by 9

10. Nine less than twice the base, b

5. _____

6. _____

7. _____

8. _____

9. _____

10. _____

a) Select a variable to represent a quantity and state what the variable represents, and b) express the second quantity in terms of the variable selected. Note: Your variable may be different from your instructor's.

11. Justin's test score was 12 points less than the average test score.

    11. a)_____

    b)_____

12. Under the new union contract, the employees' salaries will increase 5%.

    12. a)_____

    b)_____

13. Kristen and Kyle paid for their dinner separately. Their restaurant bills totaled $54.35.

    13. a)_____

    b)_____

14. The width of the rectangular garden is 5 ft less than half the length.

    14. a)_____

    b)_____

Practice Set 3.1

Write an equation to represent the problem.

15. Becky's age is 8 more than 10 times her son Jack's age. The sum of their ages is 41.

15. _____

16. In the race for president, the winner of the popular vote in each state wins all the electoral votes in that state (except in Maine and Nebraska). The president elect received 104 more than three-fourths the electoral votes of the other candidate, and there were a total of 538 electoral votes.

16. _____

17. This year's festival brought 80 more than twice the number of customers to H.H. Fortmann & Co. restaurant than last year's festival. How many customers came to the restaurant this year if the number of customers totaled 845 from both festivals?

17. _____

18. A certain beetle can pull a 2-gram object that is 150 times its own weight. How much, in grams, does the beetle weigh?

18. _____

Name:
Instructor:
Date:
Section:

3.2 Solving Application Problems

Objectives
1. Use the problem-solving procedure.
2. Set up and solve number application problems.
3. Set up and solve application problems involving money.
4. Set up and solve application problems concerning percent.

1 Use the problem-solving procedure.

Example 1 Use and display the five-step problem-solving procedure to solve the following.

There are 12 more sugar packets than imitation sweetener packets in a basket at a restaurant. If there are a total of 30 packets in all, find the number of each type of packet in the basket.

For the next four examples, set up an equation that can be used to solve the problem. Solve the equation and answer the question asked.

2 Set up and solve number application problems.

Example 2 One number is 5 less than twice another number. If the sum of the two numbers is 94, find the numbers.

Example 3 The city planning commission has researched the amount of daily traffic that travels across a one-lane bridge on Highway D. The number of cars traveling that stretch of highway in 2009 was 18 less than 3 times the number of cars in 2000. If the number of cars that traveled on the bridge in 2000 and 2009 totaled 502, find the number of cars that traveled at that location in 2009.

3 Set up and solve application problems involving money.

Example 4 Kevin Kifowit rented a carpet cleaner from a local hardware store. The store charges $30 for a 2-hour rental. How long did Kevin keep the carpet cleaner if he purchased one bottle of carpet shampoo for $12.95 and the bill totaled $102.95 before tax?

4 Set up and solve application problems concerning percent.

Example 5 The house specialty at a local restaurant is Kickin' Grilled Chicken. If a party of four each ordered the house specialty, what was the bill before tax if the total bill was $29.61 with a $6\frac{1}{2}$% sales tax?

Answers: 1. Step 1) We need to find the number of sugar and sweetener packets in the basket. 2) number of sweetener packets = x; number of sugar packets = $x + 12$; total number of packets = 30; the equation is $x + (x + 12) = 30$ 3) $x = 9$ 4) $9 + (9 + 12) = 30$ 5) There are 9 sweetener packets and 21 sugar packets in the basket. 2. 33 and 61 3. 372 cars 4. 6 h 5. $27.80

Name:
Instructor:

Date:
Section:

Practice Set 3.2

Match the statement on the left to its mathematical translation on the right.

1. The sum of twice a number and 10 is 24. ____

2. The sum of 2 consecutive integers is 11. ____

3. The sum of 2 and 11 more than a number is 24. ____

4. The sum of 2 consecutive odd integers is 24. ____

A. $x + x + 2 = 24$

B. $x + x + 2 = 11$

C. $x + 2x = 11$

D. $2x + 10 = 24$

E. $x + x + 1 = 11$

F. $2 + x + 11 = 24$

Set up an equation that can be used to solve the problem. Solve the equation and answer the question asked.

5. **Unknown Number** The sum of a number and 4 more than 3 times that number is 84. Find the number.

   5. _____

6. **House Numbers** The sum of 2 consecutive integers is 377. Find the integers.

   6. _____

7. **Even Integers** The sum of 2 consecutive even integers is 358. Find the numbers.

   7. _____

8. **Population** The town of Marthasville currently has a population of 808. If its population is increasing at a rate of 112 people per year, how long will it take for the population to reach 2264?

   8. _____

Practice Set 3.2

9. **Fax Copies** Ms. Ayres needs to fax a document from Copies R Us that charges $2.50 for the first page and $0.75 for each additional page. If Ms. Ayres' total bill was $9.25 before tax, how many pages did she fax?

9. _____

10. **Tax** The cost of Mr. Wiley's groceries after tax is $91.08. If the sales tax was $6\frac{1}{4}\%$, how much did the groceries cost before tax?

10. _____

11. **Sales Commission** Julie Soellner, a sales associate for an electronics department, receives a weekly salary of $150. She also receives a 9% commission on the total sales she makes. What must her sales be in a week if she is to make a total of $280?

11. _____

12. **Salary Increase** Jing Zhang was told that, according to her performance evaluation, her salary next year will be 6% more than her current salary. If this increase results in a new salary of $35,950, what is her current salary?

12. _____

13. **Autographs** A famous children's author was hired to sign autographs at a bookstore. She was paid $1000 plus 50 cents for each book autographed. If she received $1210 at the end of the day, find the total number of books signed by the author.

13. _____

14. **Used Car** Mr. Blossom purchased a 2006 used sedan. The cost of the car after a 5% rebate was $11,566. What was the cost of the car before the rebate?

14. _____

Name:
Instructor:
Date:
Section:

3.3 Geometric Problems

Objectives
> **1** Solve geometric problems.
>
> **Key Vocabulary**
> isosceles triangle, equilateral triangle, complementary angles, supplementary angles, vertical angles, quadrilateral, parallelogram, trapezoid, rhombus

1 Solve geometric problems.

Solve the following geometric problems.

Example 1 Isosceles Triangle In an isosceles triangle, one angle is 18° greater than the other two angles equal in measure. Find the measure of all three angles.

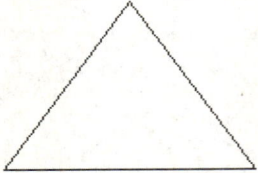

Example 2 Complementary Angles One angle is 18° less than another angle. If these angles are complementary, find the measure of each angle.

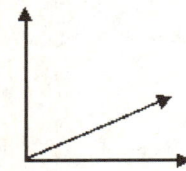

Example 3 Unknown Angles One angle of a triangle is 2° more than twice the smallest angle. The largest angle is 3° more than twice the smallest angle. Find the measures of the three angles.

Example 4 Rectangle The length of the rectangular door of an armoire is 4 inches more than 3 times its width. If the perimeter of the door is 128 inches, find the dimensions of the door.

Answers: 1. 54°, 54°, 72°   2. 36°, 54°   3. 35°, 72°, 73°   4. width = 15 in., length = 49 in.

Name: Date:
Instructor: Section:

Practice Set 3.3

Match the description on the left to its best match on the right.

1. A triangle with perimeter P _____   **A.** $A+B+C+D=360$

2. The angles of a quadrilateral _____   **B.** $2L+2W=P$

3. The angles of a triangle _____   **C.** $4s=P$

4. A rectangle with perimeter P _____   **D.** $a+b+c=P$

5. A square with perimeter P _____   **E.** $A+B+C=180$

Set up an equation that can be used to solve the problem. Solve the equation and answer the question asked.

6. **Equilateral Triangle** The perimeter of an equilateral triangle is 50.85 cm. Find the length of each side.

6. _____

7. **Complementary Angles** Two angles in a triangle are complementary angles, and one of the angles is 3 more than twice the other angle. Find the measures of the two angles.

7. _____

8. **Supplementary Angles** Two angles are supplementary, and the first angle is twice the other angle. Find the measures of the angles.

8. _____

9. **Vertical Angles** An artist is painting a mural that contains two lines crossing so that 2 pairs of vertical angles are formed. Find the measures of the angles indicated in the picture below.

9. _____

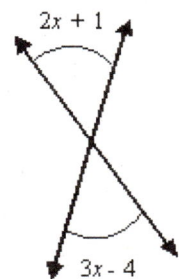

Practice Set 3.3

10. **Unknown Angles** One angle of a trapezoid has twice the measure of the smallest angle. The other two angles are equal in measure and measure 30° more than the smallest angle. Find the measures of the angles of the trapezoid.

10. _____

11. **Deck** Sam Smotkin is building a square deck. The perimeter of the deck is to be 108 ft. Determine the dimensions of the deck.

11. _____

12. **Parallelogram** Each of the two larger angles in a parallelogram is 10° less than twice the measures of the smaller angles. Find the measure of each angle.

12. _____

13. **Triangle** The largest angle in a triangle measures 20° more than the smallest angle. The third angle is 10° less than the largest angle. Find the measures of the angles.

13. _____

14. **Dimensions** The width of a rectangular table is 30 inches less than its length. If the perimeter of the table is 200 inches, find the dimensions of the table.

14. _____

Name:
Instructor:
Date:
Section:

3.4 Motion, Money, and Mixture Problems

> **Objectives**
> **1** Solve motion problems involving two rates.
> **2** Solve money problems.
> **3** Solve mixture problems.
> **Key Vocabulary**
> motion problem, money problem, mixture problem

1 Solve motion problems involving two rates.

Example 1 Distance-rate-time Complete the following tables using the fact that $d = r \cdot t$.

a)

Item	Rate	Time	Distance
Item 1	x	8	?
Item 2	$3x$	8	?

b)

Item	Rate	Time	Distance
Item 1	x	6	?
Item 2	$x - 3$	8	?

Example 2 Refer to your completed tables in Example 1. Set up the equation that will solve the problem given the following information.

a) Using the completed table in Example 1a), set up the equation if the distance traveled by item 1 and item 2 totaled 240 mi.

b) Using the completed table in Example 1b), set up the equation if item 1 and item 2 traveled the same distance.

c) Using the completed table in Example 1b), set up the equation if item 1 traveled 10 mi farther than item 2.

Example 3 Walking Sue Dubsky and her dog, Zoey, were walking on a nature trail at a rate of 3 mph. Another young hiker was hiking towards Sue and Zoey at a rate of 4 mph. After they pass each other on the trail, how long will it take for Sue and her dog to be 1 mi apart from the other hiker?

Item	Rate	Time	Distance
Sue and Zoey			
Other hiker			

Example 4 Round Trip A group of runners began running west on the Katy Trail at 4.5 mph. After a certain distance, they turned around and ran east at 4 mph back to where they met. If the runners took 0.25 hr longer running east than running west, determine the time it took them to run east.

Item	Rate	Time	Distance
West			
East			

Answers: 1a) item 1: $8x$; item 2: $24x$ b) item 1: $6x$; item 2: $8x - 24$ 2a) $8x + 24x = 240$ b) $6x = 8x - 24$ c) $6x - (8x - 24) = 10$ 3. $\frac{1}{7}$ hr or approximately 8.6 min   4. $2\frac{1}{4}$ hr

Examples 3.4

2 Solve money problems.

Example 5 Simple Interest The Schuslers invested $14,000, part at 3% simple interest and the rest at 4% simple interest for a period of 1 year. How much did they invest at each rate if their total annual interest from both accounts was $460?

Account	Principal	Rate	Interest
3% account			
4% account			

Example 6 Satellite TV Joyce Combs noticed in her December bill that her subscription rate for satellite service increased at some point from $29.95 per month to $38.95. If she paid a total of $395.40 to the satellite provider for a full calendar year, determine the month of the rate increase.

Month	Number of Months	Cost	Service Fee
Before increase			
After increase			

3 Solve mixture problems.

Example 7 Bulk Herbs The Natural Food Store sells an organic dried parsley and a generic brand dried parsley in bulk. The organic parsley sells for $7.95 per oz and the generic parsley sells for $4.95 per oz. If a 4 oz mixture costs $24.30, how much of each type of parsley was purchased?

Item	Number of Ounces	Cost per Ounce	Cost of Parsley
Organic			
Generic			
Mixture			

Example 8 Medical Research Dr. Pavia, a research scientist, ordered a 12% solution of a chemical for his supply cabinet, but received a 20% solution instead. Dr. Pavia notices he has a 10% solution in stock. How much of the 20% solution must he add to the 10% solution to make 8 liters of a 12% solution?

Solution	Number of Liters	Strength	Amount of Solution
10%			
20%			
12%			

Example 9 Cranberry Juice Cocktail A popular brand of cranberry juice cocktail consists of 18% cranberry extract. How much cranberry juice cocktail will need to be mixed with pure water to make a 12 oz mixture that is 10% cranberry extract?

Item	Ounces	Strength	Amount of Cranberry Extract
Cranberry juice cocktail			
Pure water			
Mixture			

Answers: 5. $10,000 at 3%; $4000 at 4%   6. September   7. 1.5 oz organic parsley; 2.5 oz generic parsley   8. 1.6 L   9. $6\frac{2}{3}$ oz

Name: Date:
Instructor: Section:

Practice Set 3.4

1. Complete the following table given that $d = rt$.

Item	Rate	Time	Distance
Paul walking	$x - 2$	0.5	
Paul running	x	2	

2. Complete the table given only the following information: Jacob and his brother Cody are riding their bikes. Cody travels at 10 mph while Jacob travels at 8 mph for one hour longer than Cody.

Item	Rate	Time	Distance
Jacob			
Cody			

3. Refer to your completed table in Problem 2 above. For how long did each brother travel if they each traveled the same distance?

   3. _____

4. Refer to Problem 3 above. How far did each brother travel?

   4. _____

5. **Round Trip** A retired couple takes a walk every day, making a round trip to the park and back to their house. Their rate walking to the park is 1.5 mph slower than their trip home since the trip to the park is mostly uphill. If it takes 45 min to get to the park, but 30 min to return home, how far is it from their house to the park?

   5. _____

6. **Traveling Cars** Two cars leave a jazz festival, one heading north and the other heading south. The northbound car is traveling 11 mph faster than the southbound car. If the cars are 200 miles apart after 2 hours, how fast is each car traveling?

   6. _____

7. **Rickshaw Ride** Two conference attendees rented a rickshaw while in San Diego. The rickshaw traveled at 14 mph from the hotel to the wharf, then traveled at 10 mph back to the hotel. If the trip to the wharf took 6 min less than the ride back, how far was it from the hotel to the wharf?

   7. _____

Practice Set 3.4

8. **Road Trip** Kristen's aunt drove from home to Kristen's house, a distance of 320 miles. Part of the way, Kristen's aunt drove at 70 mph, then the rest of the way she drove at 50 mph on two-lane roads. If Kristen's aunt drove for 4 hours longer at 70 mph than she did at the slower speed, how long did her aunt travel at 70 mph?

8. _____

9. **Raisin & Cashew Mixture** Wholesome Food Store sells certain foods in bulk. The cashews sell for $7.95 per pound and raisins sell for $1.95 per pound. If Helen Wiley pays $20.70, before tax, for a 6-lb mixture of cashews and raisins, how much of each item did she purchase?

9. _____

10. **Diluting Bleach** A bottle of generic bleach is 4.75% sodium hypochlorite. If $\frac{1}{4}$ cup of bleach is added to a sink filled with 2 gallons of water, find the percent of sodium hypochlorite in the sink of water. (Hint: 1 gallon = 16 cups)

10. _____

11. **Simple Interest** Hannah Cushman invested $6000, part at $3\frac{1}{2}$% simple interest and the rest at 4% for a period of 1 year.

   a) How much was invested in each account if Hannah earned a total of $220 in interest from both accounts?

   b) How much was invested at each rate if the interest earned at 4 % was $40 more than the interest earned at $3\frac{1}{2}$%?

   c) How much was invested at each rate if the interest earned at 4% was the same as the interest earned at $3\frac{1}{2}$%?

11. a)_____
    _____
    b)_____
    _____
    c)_____
    _____

Chapter 3 Vocabulary Reference Sheet

Term	Definition	Example
consecutive integers	Integers that differ by one, represented by: $x, x+1, x+2, x+3, \ldots$	Find two consecutive integers whose sum is 99. Solution: $(x)+(x+1)=99$ $2x+1=99$ $2x=98$ $x=49$ The first integer is 49 and the next is 50.
consecutive even integers	Integers that are even; they differ by two, represented by: $x, x+2, x+4, \ldots$	Find two consecutive even integers whose sum is 50. Solution: $(x)+(x+2)=50$ $2x+2=50$ $2x=48$ $x=24$ The first integer is 24 and the next is 26.
consecutive odd integers	Integers that are odd; they differ by two, represented by: $x, x+2, x+4, \ldots$	Find two consecutive odd integers whose sum is 48. Solution: $(x)+(x+2)=48$ $2x+2=48$ $2x=46$ $x=23$ The first integer is 23 and the next is 25.
isosceles triangle	A triangle with two sides that are the same length; the angles opposite the sides of equal length have equal measures	In the triangle below, angles x are each the same measure and sides y are the same length.
equilateral triangle	A triangle with three sides equal in length and three angles equal in measure	In the triangle below, the three angles all measure 60° and the three sides are the same length.
complementary angles	Two angles whose measures add to 90°	

Chapter 3 Vocabulary

supplementary angles	Two angles whose measures add to 180°	*(figure: angle split into $180-x$ and x)*
vertical angles	The opposite angles formed when two lines intersect	*(figure: two intersecting lines with angles $2x+1$ and $3x-4$)*
parallelogram	A quadrilateral that has two pairs of opposite parallel sides; opposite sides have the same length and opposite angles are equal in measure	*(figure: parallelogram with angles $2x$, x, x, $2x$)*
rhombus	A parallelogram with all four sides equal in length	*(figure: rhombus with angles x)*
quadrilateral	A four-sided figure	A rectangle, square, parallelogram, rhombus, and trapezoid are examples of quadrilaterals.
motion problem	A problem that involves two different rates and uses the formula $d = r \cdot t$	Distance traveled by Sue = $4t$ Distance traveled by Jack = $5t$ If Sue and Jack traveled a total of 12 miles, then $4t + 5t = 12$.
money problem	A problem that involves two different interest rates or two different costs	Sam invested $1000, part at 4% simple interest and the rest at 3% simple interest for 1 year. If he earned a total of $34 in interest, then $34 = 0.04x + 0.03(1000 - x)$.
mixture problem	A problem in which two or more items are being combined to form a mixture	How many liters of a 4% solution should be mixed with a 6% solution to obtain 5 liters of a 5% solution? $0.04x + 0.06(5 - x) = 0.05(5)$

Name:
Instructor:

Date:
Section:

Chapter 3 Practice Test A

1. Jim and Christi's gross monthly income is $4700. If Jim earned m dollars in one month, write an expression for Christi's monthly income.

   1._____

2. Write an expression for the number of hours in d days.

   2._____

3. The support staff's salary, s, at East Central College increased 6%. Write an expression for the support staff's new salary.

   3._____

Select a variable to represent one quantity and state what it represents. Express the second quantity in terms of the variable selected.

4. The number of pencils purchased at the college bookstore was 7 less than the number of pens purchased.

   4._____
   _____

5. The cost of one box of All Natural Macaroni and Cheese is $0.49 more than store brand macaroni and cheese.

   5._____
   _____

Set up an equation that can be used to solve the problem. Solve the problem and answer the question asked.

6. **Consecutive Page Numbers** The sum of two consecutive page numbers is 245. Find the numbers.

   6._____

7. **Math Test Scores** The score on one student's math test was 3 less than twice her best friend's math test score. Find the two scores if the sum of the scores is 126.

   7._____

8. **Tax on Groceries** Abby Suarez paid $49.02 for her groceries, which included a 4.75% sales tax. Find the cost of Abby's groceries before tax.

   8._____

Practice Test 3A

9. **Triangle** The perimeter of a triangle is 76 in. The length of the first side is twice the length of the second side, and the third side is 1 inch longer than the second side. Find the lengths of all three sides.

9. _____

10. **Triangle** The first angle of a triangle is half the measure of the second angle. The third angle is 20° less than the first angle. Find the measures of the angles.

10. _____

11. **Rectangular Window** Barb Roberts wants to hang trim around the perimeter of her rectangular window. The length of the window is 8 in. more than its width. Find the dimensions of the window if the perimeter is 64 in.

11. _____

12. **Riding Bikes** Two friends, Nate and Matthew, are 1.25 mi away from each other on opposite ends of a park. They start riding their bikes towards each other at the same time. If Nate is riding at 7 mph and Matthew is riding at 10 mph, how long will it take for them to meet?

12. _____

13. **Round Trip** It takes Johan George 8 minutes longer to drive to work during rush hour than to get home from work taking the same route. If his average speed going to work is 45 mph, and his average speed driving home is 54 mph, how far does Johan live from work?

13. _____

14. **Concert Seats** Ann and Sean, co-sponsors of a student club, paid $208 for concert tickets. They bought twice as many balcony tickets as main-floor tickets. If the balcony tickets cost $15 and the main-floor tickets cost $22, how many of each type of ticket did Ann and Sean purchase?

14. _____

15. **Acid Solution** How many gallons of a 25% acid solution must be added to 40 gallons of a 40% acid solution to get a solution that is 30% acid?

15. _____

Name:
Instructor:
Date:
Section:

Chapter 3 Practice Test B

1. A student must take 64 credit hours to graduate with an associate degree. Let c represent the number of credit hours a student has already taken towards his degree. Write an expression that represents the number of credit hours a student has left to take to graduate.

    a) $64c$ b) $64 \div c$ c) $c - 64$ d) $64 - c$

2. 18% of the residents, r, of a town own a boat. Write an expression that represents the number of residents who own a boat.

    a) $18r$ b) $0.18r$ c) $18 - r$ d) $0.18 - r$

3. The young candidate received twice as many votes, v, as the incumbent. Write an expression that represents the number of votes received by the young candidate.

    a) $2v$ b) $v + 2$ c) $0.02v$ d) $\frac{1}{2}v$

4. **Consecutive Even Integers** The sum of 3 consecutive even integers is 552. Choose the equation that can be used to find the three integers.

    a) $x + x + 1 + x + 2 = 552$ b) $x(x+1)(x+2) = 552$

    c) $x + x + 2 + x + 4 = 552$ d) $x(x+2)(x+4) = 552$

5. **Percentage Discount** A furniture store is having a sale where all merchandise is reduced 30%. If Simone Carne purchased a nightstand for $124, what was the original price of the nightstand?

    a) $37.20   b) $86.80   c) $120.28   d) $177.14

6. **Trapezoid** The largest angle of a trapezoid measures 25° more than the smallest angle. The third angle measures 10° less than the largest angle. The fourth angle measures 20° more than the smallest angle. Find the measure of the smallest angle.

    a) 40°   b) 55°   c) 75°   d) 100°

7. **Sales Tax** Fran McDoran paid $28.52 for a DVD set, which included a 6.25% sales tax. Find the cost of the DVD set before tax.

    a) $17.83   b) $26.84   c) $27.83   d) $30.30

8. **Wagon and a Bike** David White begins pulling his son in a wagon at 2 mph. David's daughter begins riding her bike alongside the wagon and is traveling at 5 mph. How long will it take for David and his daughter to be 0.5 mi apart?

    a) 4 min   b) 7 min   c) 10 min   d) 16 min

Practice Test 3B

9. **Selling Spice Packets** Kendall Mauer sold two kinds of spice packets at last year's festival and made a gross profit of $484.50. If she sold a total of 153 packets, some for $2.25 and the rest for $5, how many of the $5-packets did she sell?

   a) 51 b) 71 c) 92 d) 102

10. **Simple Interest** Francis Bose invested $5000, part at 3% simple interest and the rest at 5% simple interest for a period of 1 year. If he received a total annual interest of $214 from both investments, how much did he invest at 5%?

    a) $1500 b) $1800 c) $3200 d) $3500

11. **Triangle** One angle of a triangle is 20° less than the largest angle. If the third angle is 10° more than the smallest angle, find the measure of the smallest angle.

    a) 40° b) 50° c) 60° d) 70°

12. **Store Coupons** The department store Kitchens and Things has two different coupons. One coupon offers $5 off any purchase. The other coupon offers 20% off any purchase. How much would someone need to spend at Kitchens and Things so that both coupons result in the same discount?

    a) $15 b) $18 c) $20 d) $25

13. **Making Meatballs** A recipe for Italian meatballs calls for a mixture of pork and beef. The store carries ground beef at $2.99 per pound, and ground pork at $3.99 per pound. How many pounds of beef could be bought to make 10 lb of a beef-and-pork mixture that costs $3.75 per pound?

    a) 2.4 lb b) 4 lb c) 7.6 lb d) 8 lb

14. **Round Trip** Morgan Taylor's mother drove to her university to see Morgan perform in a school play. It took her mother 30 minutes longer to drive to the play at 50 mph than it did to drive home at 55 mph. How long did it take Morgan's mother to get home?

    a) 5 h b) $5\frac{1}{2}$ h c) 6 h d) $6\frac{1}{2}$ h

15. **Semester Grades** Jon Lee's grade point average (on a 4-point scale) this semester was 1.026 points more than last semester's grade point average. If the sum of his grade point averages is 3.4, what is his current grade point average?

    a) 3.314 b) 2.374 c) 2.213 d) 1.187

Name:
Instructor:

Date:
Section:

4.1 Exponents

Objectives
1. Review exponents.
2. Learn the rules of exponents.
3. Simplify an expression before using the expanded power rule.

Key Vocabulary
base, exponent, product rule, quotient rule, zero exponent rule, power rule, power of a product rule, power of a quotient rule, expanded power rule

1 Review exponents.

Example 1

a) In the expression $a^4b^2c^3$, what is the numerical coefficient?

b) In the expression $4st^2$, what is the exponent of the variable s?

c) Show the expanded form of m^3n^4.

2 Learn the rules of exponents.

Example 2 Use the product rule for exponents to simplify.

a) $b^8 \cdot b^3$ b) $m^2 \cdot m$ c) $c^4 \cdot d^2$ d) $3^3 \cdot 3^3$

Example 3 Use the quotient rule for exponents to simplify.

a) $\dfrac{r^6}{r^2}$ b) $\dfrac{h^9}{h}$ c) $\dfrac{t^5}{t^5}$ d) $\dfrac{g^7}{g^{10}}$

Example 4 Use the zero exponent rule to simplify.

a) 9^0 b) $(4c)^0$ c) $6(df)^0$ d) -2^0

Answers: 1a) 1 b) 1 c) $m \cdot m \cdot m \cdot n \cdot n \cdot n \cdot n$ 2a) b^{11} b) m^3 c) c^4d^2 d) 3^6 3a) r^4 b) h^8 c) 1 d) g^{-3} or $\dfrac{1}{g^3}$ 4a) 1 b) 1 c) 6 d) -1

Examples 4.1

Example 5 Use the power rule for exponents to simplify.

a) $(p^3)^4$ b) $(2^3)^2$ c) $(g^0)^2$ d) $(k^4)^0$

Example 6 Use the expanded power rule for exponents to simplify.

a) $(3bc^2)^2$ b) $(-5m^3n)^3$ c) $\left(-\dfrac{1}{z}\right)^5$

d) $\left(\dfrac{16f^3g^4}{8g}\right)^3$ e) $\left(\dfrac{-3p^4r}{18p^4r^3}\right)^2$ f) $(5b^3c^4)^2(4b^2c^2)^2$

3 Simplify an expression before using the expanded power rule.

Example 7 Simplify the expressions in Example 6 by first simplifying within the parentheses.

a) $(3bc^2)^2$ b) $(-5m^3n)^3$ c) $\left(-\dfrac{1}{z}\right)^5$

d) $\left(\dfrac{16f^3g^4}{8g}\right)^3$ e) $\left(\dfrac{-3p^4r}{18p^4r^3}\right)^2$ f) $(5b^3c^4)(4b^2c^2)^2$

Answers: 5a) p^{12} b) 64 c) 1 d) 1 6a) $9b^2c^4$ b) $-125m^9n^3$ c) $-\dfrac{1}{z^5}$ d) $8f^9g^9$ e) $\dfrac{1}{36r^4}$ f) $400b^{10}c^{12}$ 7. Same answers as Example 6

Name:
Instructor:

Date:
Section:

Practice Set 4.1

Use the choices below to fill in each blank.

    base   exponent   add   subtract   multiply

    denominator   numerator   one   zero   coefficient

    power   factor   expression   simplify

1. In the expression $9r^2$, 9 is the _____.

2. In the expression m^5, m is the _____.

3. When multiplying expressions with the same base, keep the base and _____ the exponents.

4. Any real number, except 0, raised to the _____ power equals one.

5. When dividing expressions with the same base, keep the base and subtract the exponent in the _____ from the exponent in the _____.

6. When raising an exponential expression to a power, keep the base and _____ the exponents.

7. The expanded power rule states that every _____ within parentheses is raised to the power outside the parentheses.

Rewrite using exponents.

8. $3 \cdot 3 \cdot 3 \cdot 3 \cdot x \cdot x \cdot y \cdot y \cdot y \cdot z$ 8. _____

9. $2x \cdot 2x \cdot 2x \cdot 2x \cdot 2x$ 9. _____

10. $-a \cdot -a \cdot -a$ 10. _____

Use the rules for exponents to simplify.

11. $w^2 \cdot w^4 \cdot w^6$ 12. $\dfrac{100r^5}{4r}$ 11. _____

12. _____

13. $b(2b)(3b^2)$ 14. $\dfrac{-m^3 n}{m^6 n^3}$ 13. _____

14. _____

Practice Set 4.1

15. $\left(\dfrac{36tv^3}{4t^4v^2}\right)^2$

16. $(100x^2b^3)\cdot(x^2b)^3$

17. $(35)^2(-2s^4)^3$

18. $\left(\dfrac{12y^6z^3}{10xy^2}\right)^2$

19. $(-2c^3d^2)^3(-3bc^2d)^2$

20. $(1.2f^6)^2$

15. _____
16. _____
17. _____
18. _____
19. _____
20. _____

Problem Solving

21. What is the volume of a cube with edge of length $3x^2y^4$ cm?

21. _____

22. What is the area of the triangle with base $4s^3$ in. and height $8s^2$ in.?

22. _____

Name:
Instructor:

Date:
Section:

4.2 Negative Exponents

Objectives
1. Understand the negative exponent rule.
2. Simplify expressions containing negative exponents.

Key Vocabulary
negative exponent rule, fraction raised to a negative exponent rule

1 Understand the negative exponent rule.

Example 1 Simplify by the quotient rule. Write answers with negative exponents.

a) $\dfrac{w^2}{w^6}$ b) $\dfrac{6}{6^3}$ c) $\dfrac{t}{t^2}$ d) $\dfrac{4^3}{4^5}$

Example 2 Simplify by dividing out common factors. Write answers with negative exponents.

a) $\dfrac{s}{s^2}$ b) $\dfrac{7^4}{7^5}$ c) $\dfrac{z^3}{z^5}$ d) $\dfrac{4^3}{4^5}$

2 Simplify expressions containing negative exponents.

Example 3 Use the negative exponent rule to write each expression with a positive exponent and simplify.

a) p^{-2} b) $(-4)^{-3}$ c) $\dfrac{1}{r^{-2}}$ d) $\dfrac{1}{2^{-3}}$

e) $(v^{-3})^5$ f) $(6^{-2})^2$ g) $m^5 \cdot m^{-8}$ h) $4^{-1} \cdot 4^{-2}$

i) $\dfrac{2^{-5}}{2^{-2}}$ j) $\dfrac{a^{-4}}{a^6}$ k) $(-4g^{-3})^{-2}$ l) $-\dfrac{5k^3 t^2}{15k^3 t^{-2}}$

Answers: 1a) w^{-4} b) 6^{-2} c) t^{-1} d) 4^{-2} 2a) $\dfrac{1}{s}=s^{-1}$ b) $\dfrac{1}{7}=7^{-1}$ c) $\dfrac{1}{z^2}=z^{-2}$ d) $\dfrac{1}{4^2}=4^{-2}$ 3a) $\dfrac{1}{p^2}$ b) $-\dfrac{1}{4^3}=-\dfrac{1}{64}$
c) r^2 d) $2^3=8$ e) $v^{-15}=\dfrac{1}{v^{15}}$ f) $6^{-4}=\dfrac{1}{6^4}=\dfrac{1}{1296}$ g) $m^{-3}=\dfrac{1}{m^3}$ h) $4^{-3}=\dfrac{1}{4^3}=\dfrac{1}{64}$ i) $2^{-3}=\dfrac{1}{2^3}=\dfrac{1}{8}$ j) $a^{-10}=\dfrac{1}{a^{10}}$
k) $\dfrac{g^6}{16}$ l) $-\dfrac{t^4}{3}$

Examples 4.2

Example 4 Simplify each fraction raised to a negative exponent.

a) $\left(\dfrac{3}{8}\right)^{-2}$

b) $\left(\dfrac{n^3}{p^5}\right)^{-4}$

c) $\left(\dfrac{-2u^4v^{-1}}{7u^7v^2}\right)^2$

d) $\left(\dfrac{5a^{-2}b^{-3}c^4}{6b^{-1}c^{-2}d^3}\right)^{-2}$

Example 5 Evaluate.

a) $2^{-1} \cdot 3^{-2}$

b) $5^{-1} + 3^{-1}$

c) $4 - 2^{-1} \cdot 3^{-1}$

d) $\left(\dfrac{1}{2}\right)^{-2} + 3^2$

Answers: 4a) $\dfrac{64}{9}$ b) $\dfrac{p^{20}}{n^{12}}$ c) $\dfrac{4}{49u^6v^6}$ d) $\dfrac{36a^4b^4d^6}{25c^{12}}$ 5a) $\dfrac{1}{18}$ b) $\dfrac{8}{15}$ c) $3\dfrac{5}{6}$ d) 13

Name:
Instructor:

Date:
Section:

Practice Set 4.2

Answer True or False (if false, write correct answer).

1. $3^{-3} = \dfrac{1}{3 \cdot 3} = \dfrac{1}{9}$

2. $\left(\dfrac{m}{n}\right)^{-1} = \dfrac{n}{m}$

3. $r^{-2} = \dfrac{1}{r^2}$

4. $\left(\dfrac{1}{2}\right)^{-3} = 2 \cdot -3 = -6$

5. $(-5)^{-2} = \left(-\dfrac{1}{5}\right)\left(-\dfrac{1}{5}\right) = \dfrac{1}{25}$

6. $\dfrac{t^2 u^{-3}}{tu} = tu^2$

7. $\left(6^{-2} + 2^{-2}\right)^0 = 2$

8. $3^{-2} + 3^2 = 0$

9. $-x^{-2} = -\dfrac{1}{x^2}$

10. $\dfrac{-w^{-4}}{w^{-4}} = -1$

1. _____
2. _____
3. _____
4. _____
5. _____
6. _____
7. _____
8. _____
9. _____
10. _____

Simplify and rewrite each with no negative exponents.

11. $m^{-3} n^2$

12. $\left(5c^{-1} d\right)^{-2}$

13. $\left(\dfrac{-3t^3}{7t^3}\right)^{-2}$

14. $\left(4^{-2}\right)\left(8\right)^0$

15. $\left(\dfrac{-5^{-1}}{3}\right)^2$

16. $6^{-1} + 2^{-2}$

11. _____
12. _____
13. _____
14. _____
15. _____
16. _____

Practice Set 4.2

What should the value of n be to make the following true?

17. $4^n = \dfrac{1}{16}$

17. _____

18. $\left(\dfrac{3}{2}\right)^n = \dfrac{27}{8}$

18. _____

19. $\left(-\dfrac{1}{5}\right)^n = -125$

19. _____

20. $\left(\dfrac{a^n b^3}{c^n}\right)^{-3} = \dfrac{a^{12}}{c^{12} b^9}$

20. _____

Name: Date:
Instructor: Section:

4.3 Scientific Notation

> **Objectives**
> 1. Convert numbers to and from scientific notation.
> 2. Recognize numbers in scientific notation with a coefficient of 1.
> 3. Do calculations using scientific notation.
>
> **Key Vocabulary**
> scientific notation

1 Convert numbers to and from scientific notation.

Example 1 Write the following numbers in scientific notation.

a) 4,320,000   b) 0.000512   c) 8,000,000,000   d) 0.002002

Example 2 Write each number in decimal form (without exponents).

a) 5.24×10^7 b) 6.03×10^{-5} c) 8.0×10^{-4} d) 4.37×10

2 Recognize numbers in scientific notation with a coefficient of 1.

Example 3 Write each quantity without the given metric prefix.

a) 76 megabytes   b) 3 micrograms   c) 492 kilowatts   d) 15 nanoseconds

3 Do calculations using scientific notation.

Example 4 Multiply. Write the answer in decimal form.

a) $(3.8 \times 10^4)(2.2 \times 10^{-2})$ b) $(1.6 \times 10^{-3})(5.0 \times 10^6)$ c) $(6.25 \times 10^2)(8.0 \times 10^2)$

Example 5 Divide. Write the answer in decimal form.

a) $\dfrac{2.5 \times 10^6}{1.25 \times 10^3}$ b) $\dfrac{1.6 \times 10^{-3}}{2.0 \times 10^{-5}}$ c) $\dfrac{175,000}{0.00025}$

Answers: 1a) 4.32×10^6 b) 5.12×10^{-4} c) 8.0×10^9 d) 2.002×10^{-3} 2a) 52,400,000 b) 0.0000603 c) 0.0008 d) 43.7
3a) 76,000,000 bytes b) 0.000003 g c) 492,000 watts d) 0.000000015 sec 4a) 836 b) 8000 c) 500,000 5a) 2000
b) 80 c) 700,000,000

Name: Date:
Instructor: Section:

Practice Set 4.3

Match each metric prefix with its decimal number.

1. mega   A. 0.000000001   1. _____
2. nano   B. 0.000001   2. _____
3. milli   C. 0.001   3. _____
4. giga   D. 1000   4. _____
5. micro   E. 1,000,000   5. _____
6. kilo   F. 1,000,000,000   6. _____

Write the decimal number value.

7. 2.39×10^4 8. 8.6×10^{-3} 7. _____

8. _____

9. 7.5×10^0 10. 1.3×10^9 9. _____

10. _____

Perform the indicated calculations using scientific notation. Keep answers in scientific notation.

11. $(6.0 \times 10^9)(4.2 \times 10^{-3})$ 11. _____

12. $(8.1 \times 10^{-4})(1.3 \times 10^{-2})$ 12. _____

13. $(2.7 \times 10)(5.0 \times 10^4)$ 13. _____

14. $(0.0009)(4500)$ 14. _____

15. $\dfrac{1.4 \times 10^{-3}}{2.8 \times 10}$ 15. _____

16. $\dfrac{9.9 \times 10^2}{3.3 \times 10^{-2}}$ 16. _____

Practice Set 4.3

17. $\dfrac{1.5 \times 10^4}{7.5 \times 10^4}$

17. _____

18. $\dfrac{0.6}{12,000}$

18. _____

Solve each problem. Write answers in scientific notation.

19. A typical 500 megawatt coal power plant produces 3.5 billion kWh per year. That is enough energy for 4 million of our incandescent light bulbs to operate year round. How many kWh are needed for one light bulb?

19. _____

20. On a digital channel each broadcaster sends a 19.39 megabits-per-second (mbps) stream of digital data. How many bits are sent per hour?

20. _____

Name:
Instructor:

Date:
Section:

4.4 Addition and Subtraction of Polynomials

Objectives
1. Identify polynomials.
2. Add polynomials.
3. Subtract polynomials.
4. Subtract polynomials in columns.

Key Vocabulary
polynomial, descending order, monomial, binomial, trinomial, degree of a term, degree of a polynomial

1 Identify polynomials.

Example 1 State whether each expression is a monomial, binomial, or trinomial.

a) $8w^2 + 6w - 1$ b) $3xy^2$ c) $\frac{2}{3}x + \frac{1}{4}$ d) $7c^2b - 2a$

Example 2 State the degree of each polynomial.

a) $9k^2 - 7k^3$ b) $4s^3t^2 + 6t^4$ c) $w + 1$ d) $8p^2 + 7pr^2 - 5$

2 Add polynomials.

Example 3 Add the following polynomials using columns. (Remember to arrange in descending order first.)

a) $(6t^2 - 4t + 3) + (9 + 7t - 2t^2)$

b) $(32m^2 - 8mn + 9n^2) + (3m^2 + 2mn)$

c) $(-p^2 - 2r^2) + (5rp + \frac{1}{2}r^2 + 3p^2)$

d) $(6.3w^3 - 2.1w^2 + 1.9w + 7) + (4.2w - 5.3w^2)$

Answers: 1a) trinomial b) monomial c) binomial d) binomial 2a) 3 b) 5 c) 1 d) 3 3a) $4t^2 + 3t + 12$ b) $35m^2 - 6mn + 9n^2$ c) $-\frac{3}{2}r^2 + 5rp + 2p^2$ d) $6.3w^3 - 7.4w^2 + 6.1w + 7$

Examples 4.4

3 Subtract polynomials.

Example 4 Subtract the following polynomials. Write answers in descending order.

a) $(6x^2 - 3x + 2) - (9x^2 + 11x - 7)$

b) $(4c^3 - 1) - (2c^2 + 5c - 1)$

c) $\left(8m^2 + 10m - \dfrac{5}{8}\right) - \left(\dfrac{3}{4}m^2 - 2 + 6m\right)$

d) Subtract $(9 - t^2)$ from $(4t^3 + 7 - t + 6t^2)$.

4 Subtract polynomials in columns.

Example 5 Subtract the following polynomials using columns.

a) $(4p^2 + 7p - 9) - (-2p^2 - p + 3)$

b) $(-2u - 3u^2) - (5 + 8u^2)$

c) $(10w^3 - 4w + 1) - (w^3 + 2w^2 - 9w + 6)$

d) Subtract $(-8n + 7)$ from $(3n^2 - 5n - 2)$.

Answers: 4a) $-3x^2 - 14x + 9$ b) $4c^3 - 2c^2 - 5c$ c) $\dfrac{29}{4}m^2 + 4m + \dfrac{11}{8}$ d) $4t^3 + 7t^2 - t - 2$ 5a) $6p^2 + 8p - 12$
b) $-11u^2 - 2u - 5$ c) $9w^3 - 2w^2 + 5w - 5$ d) $3n^2 + 3n - 9$

Name:
Instructor:

Date:
Section:

Practice Set 4.4

Complete the following chart.

	Polynomial	Number of terms	Rewrite in descending order	State the degree
1.	$-3m$			
2.	$t^2 - 4t^3 + 1$			
3.	$\frac{1}{2}p + p^3 - 7p^2 - 9$			
4.	$-\frac{1}{5} - 2c^3 + 0.8c^4$			
5.	$6 + p^5$			
6.	-5			

Perform the indicated operation (write answers in descending order).

7. $(9x^2 + 7x + 5) + (-x^2 - 3)$

7. _____

8. $(7x^3 - 2x + 1) - (7x^3 - 2x - 1)$

8. _____

9. $\left(\frac{1}{2}y^2 - \frac{1}{4}\right) + \left(\frac{1}{2}y^2 + \frac{1}{4}\right)$

9. _____

10. $y - (y^2 - 3y + 2)$

10. _____

11. $(3.2z + 1.8z^2 - 9) + (-4.2z^2)$

11. _____

12. $(9.3c^2 - 7.4c) + (3.9c^2) - (-1.2c)$

12. _____

Practice Set 4.4

Add or subtract using columns.

13. $(2x^2 + 3) + (4x + 7) + (6x^2 - 5x)$

13. _____

14. $t + (t^3 - 2) + (t^2 - 8t + 11)$

14. _____

15. $(-a^2 - 2a - 3) - (-4a - 5)$

15. _____

16. $(r + 2r^2 - 7) - (6 + 3r)$

16. _____

17. Subtract $(-11p - 6)$ from $(8p^2 - 7p + 3)$.

17. _____

18. Subtract $(5k + 2)$ from $(9k^2 - 13)$.

18. _____

19. What is the perimeter of a triangle with sides measuring $(a + 2)$ cm, $(6a - 7)$ cm, and $(3a)$ cm?

19. _____

20. A board measuring $(7z - 3)$ m long is cut into two pieces. If the first piece is $(2z + 1)$ m in length, how long is the second piece?

20. _____

Name:
Instructor:
Date:
Section:

4.5 Multiplication of Polynomials

Objectives
1. Multiply a monomial by a monomial.
2. Multiply a polynomial by a monomial.
3. Multiply binomials using the distributive property.
4. Multiply binomials using the FOIL method.
5. Multiply binomials using formulas for special products.
6. Multiply any two polynomials.

Key Vocabulary
FOIL method, product of sum and difference, difference of two squares, square of a binomial

1 Multiply a monomial by a monomial.

Example 1 Multiply.

a) $\left(3t^3\right)\left(\dfrac{1}{3}t^2\right)$ b) $\left(-8p^4r^2\right)(pr)$ c) $(1.4m)(0.5n)$

2 Multiply a polynomial by a monomial.

Example 2 Multiply.

a) $7v^3\left(v^2-4\right)$ b) $-2b^2c^3(-b+5c)$ c) $\left(a^3+\dfrac{1}{3}\right)6a$

3 Multiply binomials using the distributive property.

Example 3 Multiply (use the distributive property to write in expanded form).

a) $(20+9)(10+5)$ b) $\left(1+\dfrac{1}{4}\right)\left(4-\dfrac{1}{2}\right)$ c) $(6+0.2)(3+0.5)$

4 Multiply binomials using the FOIL method.

Example 4 Multiply.

a) $(4w-3)(5w+7)$ b) $(6f+11)(3f-11)$ c) $(z-9)(4-z)$

Answers: 1a) t^5 b) $-8p^5r^3$ c) $0.7mn$ 2a) $7v^5-28v^3$ b) $2b^3c^3-10b^2c^4$ c) $6a^4+2a$ 3a) $200+90+100+45=435$ b) $4+1-\dfrac{1}{2}-\dfrac{1}{8}=4\dfrac{3}{8}$ c) $18+0.6+3+0.1=21.7$ 4a) $20w^2+13w-21$ b) $18f^2-33f-121$ c) $-z^2+13z-36$

Examples 4.5

5 Multiply binomials using formulas for special products.

Example 5 Multiply.

a) $(6t + 5r)(6t - 5r)$ b) $(10v + 0.7)^2$ c) $\left(4p - \dfrac{1}{2}\right)^2$

6 Multiply any two polynomials.

Example 6 Multiply.

a) $(3b - 7)(2b^2 - b + 4)$ b) $(5m^3 - 7m^2 + 3m + 9)(m - 1)$

Answers: 5a) $36t^2 - 25r^2$ b) $100v^2 + 14v + 0.49$ c) $16p^2 - 4p + \dfrac{1}{4}$ 6a) $6b^3 - 17b^2 + 19b - 28$
b) $5m^4 - 12m^3 + 10m^2 + 6m - 9$

Name:
Instructor:

Date:
Section:

Practice Set 4.5

Match the expression on the left to its simplified form on the right.

1. $ab(ab)$
2. $a(ab) + b(ab)$
3. $(a+b)(a-b)$
4. $(a+b)^2$
5. $(-a-b)^2$
6. $(a+b)^3$
7. $(-a+b)(-a-b)$
8. $-(a-b)^2$

A. $a^2 - b^2$
B. $-a^2 + 2ab - b^2$
C. $a^2 b^2$
D. $a^3 + 3a^2 b + 3ab^2 + b^3$
E. $a^2 + 2ab + b^2$
F. $a^2 b + ab^2$
G. $-a^2 - b^2$
H. $a^3 + b^3$

1. _____
2. _____
3. _____
4. _____
5. _____
6. _____
7. _____
8. _____

Multiply.

9. $(30m^3 n)\left(-\dfrac{1}{6} mn^2\right)$

9. _____

10. $(0.4c^3)\left(\dfrac{2}{5} c\right)$

10. _____

11. $9t^2 v\left(t^3 - v^2\right)$

11. _____

12. $(z^2 - z + 8)3z$

12. _____

Angel/Runde, *Elementary Algebra for College Students*, 9e

Practice Set 4.5

13. $(7d+4f)(2d-9f)$ 13. _____

14. $(6g-5h)(h-g)$ 14. _____

15. $\left(2w+\dfrac{1}{3}\right)\left(2w-\dfrac{1}{3}\right)$ 15. _____

16. $(5p-3r)^2$ 16. _____

17. $(3b+4)(8b^2-7b+6)$ 17. _____

18. $\left(9k^2-5k-3\right)\left(k-\dfrac{1}{3}\right)$ 18. _____

19. Find the area of a trapezoid with one base of $(3s-4)$ in., another base of $(7s+8)$ in., and a height of $(6-s)$ in.   19. _____

Name:
Instructor:

Date:
Section:

4.6 Division of Polynomials

Objectives
1. Divide a polynomial by a monomial.
2. Divide a polynomial by a binomial.
3. Check division of polynomial problems.
4. Write polynomials in descending order when dividing.

Key Vocabulary
divisor, dividend, quotient, remainder

1 Divide a polynomial by a monomial.

Example 1 Divide.

a) $\dfrac{10x^2 + 5x + 1}{5}$ b) $\dfrac{-12c^3 + 18c^2 - 6c}{6c}$ c) $\dfrac{49v^5 - 14v^4 + 21v^3 - 35v^2}{-7v^2}$

2 Divide a polynomial by a binomial.

Example 2 Divide.

a) $\dfrac{3t^2 - t - 14}{t + 2}$ b) $\dfrac{28p^2 - 27p + 5}{4p - 1}$ c) $\dfrac{14k^2 - 11k - 15}{2k - 3}$

3 Check division of polynomial problems.

Example 3 Use (divisor × quotient) + remainder = dividend to decide if each division statement is true or false.

a) $4x - 3 \overline{\smash{)}28x^2 - 13x + 6}$ with quotient $7x - 2$

b) $2x - 3 \overline{\smash{)}16x^2 - 14x - 12}$ with quotient $8x + 5 + \dfrac{3}{2x - 3}$

4 Write polynomials in descending order when dividing.

Example 4 Divide.

a) $(100n^2 - 49) \div (10n + 7)$ b) $(64b^3 + 125) \div (4b + 5)$ c) $\dfrac{2x + 3x^3 - 4}{-1 + x}$ d) $\dfrac{7 - 4x + 9x^2 - 5x^3}{x - 1}$

Answers: 1a) $2x^2 + x + \dfrac{1}{5}$ b) $-2c^2 + 3c - 1$ c) $-7v^3 + 2v^2 - 3v + 5$ 2a) $3t - 7$ b) $7p - 5$ c) $7k + 5$ 3a) False b) True

4a) $10n - 7$ b) $16b^2 - 20b + 25$ c) $3x^2 + 3x + 5 + \dfrac{1}{x - 1}$ d) $-5x^2 + 4x + \dfrac{7}{x - 1}$

Name:
Instructor:

Date:
Section:

Practice Set 4.6

Using the following division problem, answer questions 1 through 4.

$$3x-2 \overline{\smash{\big)}\, 12x^2 + 7x - 12} \quad \rightarrow \quad 4x + 5 - \dfrac{2}{3x-2}$$

1. What is the divisor?   1. _____

2. What is the dividend?   2. _____

3. What is the quotient?   3. _____

4. What is the remainder?   4. _____

5. Divide $54c + 18$ by 9.   5. _____

6. Divide $72k^5 + 16k^3 + 56$ by $-8k^2$.   6. _____

7. Divide $9t^3 - 7t^2 - 6t + 5$ by $3t$.   7. _____

8. Divide $-18d + 6d^3 - 12 + 24d^2$ by $6d$.   8. _____

Perform each division.

9. $(10x^2 + 11x + 3) \div (5x + 3)$ 9. _____

10. $(2y^3 + 7y^2 + 6y - 3) \div (2y + 3)$ 10. _____

11. $(-p - 21 + 2p^3) \div (2p - 7)$ 11. _____

Practice Set 4.6

12. $(-13m - 4 + 9m^3) \div (3m+1)$

12. _____

13. $\dfrac{6a^3 + a^2 + 2a + 1}{3a - 1}$

13. _____

14. $\dfrac{3f^3 - 4f^2 + 2f + 3}{f - 3}$

14. _____

15. $\dfrac{64h^2 - 100}{8h + 10}$

15. _____

16. $\dfrac{121 - 36n}{11 - 6n}$

16. _____

17. $\dfrac{216s^3 - 343}{6s - 7}$

17. _____

18. $\dfrac{r^3 + 1000}{r^2 - 10r + 100}$

18. _____

Chapter 4 Vocabulary Reference Sheet

Term	Definition	Example
base	In $a^b = n$, a is the base.	In z^3, z is the base.
exponent	In $a^b = n$, b is the exponent.	In z^3, 3 is the exponent.
product rule	$a^m \cdot a^n = a^{m+n}$	$h^3 \cdot h^2 = h^5$
quotient rule	$\dfrac{a^m}{a^n} = a^{m-n}$, $a \neq 0$	$\dfrac{h^3}{h} = h^{3-1} = h^2$
zero exponent rule	$a^0 = 1$, $a \neq 0$	$c^0 = 1$ or $(8p^4)^0 = 1$
power rule	$(a^m)^n = a^{m \cdot n}$	$(r^3)^4 = r^{12}$
power of a product rule	$(ab)^m = a^m b^m$	$(3f)^3 = 3^3 f^3 = 27 f^3$
power of a quotient rule	$\left(\dfrac{a}{b}\right)^m = \dfrac{a^m}{b^m}$, $b \neq 0$	$\left(\dfrac{2}{5}\right)^4 = \dfrac{2^4}{5^4} = \dfrac{16}{625}$
expanded power rule	$\left(\dfrac{ax}{by}\right)^m = \dfrac{a^m x^m}{b^m y^m}$, $b \neq 0$ and $y \neq 0$	$\left(\dfrac{2r^2 s}{t^4}\right)^3 = \dfrac{2^3 (r^2)^3 s^3}{(t^4)^3} = \dfrac{8 r^6 s^3}{t^{12}}$
negative exponent rule	$a^{-m} = \dfrac{1}{a^m}$, $a \neq 0$	$w^{-6} = \dfrac{1}{w^6}$
fraction raised to a negative exponent rule	$\left(\dfrac{a}{b}\right)^{-m} = \left(\dfrac{b}{a}\right)^m$, $a \neq 0$ and $b \neq 0$	$\left(\dfrac{4}{5}\right)^{-3} = \left(\dfrac{5}{4}\right)^3 = \dfrac{5^3}{4^3} = \dfrac{125}{64}$
scientific notation	A number written in scientific notation has the form $a \times 10^n$, where $1 \leq a < 10$ and n is an integer.	$52{,}000 = 5.2 \times 10^4$
polynomial	An expression containing the sum of a finite number of terms of the form ax^n, for any real number a and any whole number n	$7m^3 - 6m + 5$ or $\dfrac{2}{3} r^2 + 5$
monomial	A polynomial with one term	$-6t^3$
binomial	A polynomial with two terms	$-2r^2 + 3$
trinomial	A polynomial with three terms	$4p^2 - 3p + 2$

Chapter 4 Vocabulary

degree of a term	The exponent of the variable in that term	$8b^3$ has a degree of 3. Note: A constant has a degree of 0; for example, the degree of 6 is 0. Note: If a term has more than one variable, the degree of the term is the sum of the degrees of the variables; for example, the degree of $6x^2y^3$ is 5.
degree of a polynomial	The same as that of its highest degree term	$7w^3 + 4w^2 - 1$ is a 3rd degree polynomial.
descending order	When the exponents on the variables decrease from left to right	$9v^2 - 8v + 7$ is written in descending order.
FOIL method	Multiply the first terms (F), outer terms (O), inner terms (I), and last terms (L) of two binomials.	$(u-4)(u+5) = u^2 + 5u - 4u - 20$ $= u^2 + u - 20$
product of sum and difference	$(a+b)(a-b) = a^2 - b^2$. The product $a^2 - b^2$ is called **the difference of two squares.**	$(n-2)(n+2) = n^2 - 4$
square of a binomial	$(a+b)^2 = a^2 + 2ab + b^2$ $(a-b)^2 = a^2 - 2ab + b^2$	$(g+5)^2 = g^2 + 10g + 25$ $(g-3)^2 = g^2 - 6g + 9$
dividend	The value or expression to be divided	In $(k^2 + 2k - 3) \div (k-1)$, $(k^2 + 2k - 3)$ is the dividend.
divisor	The value or expression dividing by	In $(k^2 + 2k - 3) \div (k-1)$, $(k-1)$ is the divisor.
quotient	The value or expression obtained as a result of division	In $(k^2 + 2k - 3) \div (k-1)$, $(k+3)$ is the quotient.
remainder	The value or expression remaining after division	In $(k^2 + 2k - 3) \div (k-1)$, there is no remainder (the remainder is 0).

Name:
Instructor:

Date:
Section:

Chapter 4 Practice Test A

Simplify.

1. $m^5 \cdot m^3$

    1. _____

2. $(-c^3)^6$

    2. _____

3. $\left(-\dfrac{3t^2 s}{r^3}\right)^3$

    3. _____

4. $\left(\dfrac{14a^3 b}{7ab^2}\right)^2$

    4. _____

5. 4^{-3}

    5. _____

6. $\dfrac{1}{m^{-4}}$

    6. _____

7. $(5p^4 r^{-1})(6p^{-2} r^6)$

    7. _____

8. $\left(\dfrac{15u^{-2} v}{5uv^{-3}}\right)^{-1}$

    8. _____

Write each of the following as a decimal number (without exponents) without the metric prefix.

9. 13.2 nanoseconds

10. 9.7 gigabytes

    9. _____

    10. _____

Perform each indicated operation and write your answer as a decimal number.

11. $(2.6 \times 10^2)(5.0 \times 10^{-4})$

12. $\dfrac{1.8 \times 10^{-2}}{3.0 \times 10^3}$

    11. _____

    12. _____

Practice Test 4A

Convert each number to scientific notation, then calculate, and write your answer in scientific notation.

13. $\dfrac{650,000}{0.025}$

14. $(0.00003)(4,200,000)$

13. _____

14. _____

Perform the indicated operations.

15. $(-3n^2 + 4n - 5) + (2n^2 - n + 7)$

15. _____

16. $(3.5w - 1.2) + (-2w + 1.3)$

16. _____

17. Subtract $(7y + 8)$ from $(-9y - 3)$.

17. _____

18. $3(4t^2 - 3t + 2) - (6t^2 + 1)$

18. _____

19. $(3a - 3)(2a + 2)$

19. _____

20. $-2f(11f^2 + 5f - 9)$

20. _____

21. $\dfrac{4g^2 - 32g + 16}{4g}$

21. _____

22. $\dfrac{2x^2 - x - 6}{2x + 3}$

22. _____

23. The Moon is so close to Earth that it appears to be the same size as the Sun; however, the diameter of the Sun is approximately 8.64×10^5 miles, which is approximately 400 times the diameter of the Moon. What is the approximate diameter of the Moon (in miles)?

23. _____

24. Find the perimeter of an equilateral triangle with each side measuring $(6x - 1)$ in.

24. _____

Chapter 4 Practice Test B

Simplify.

1. $p^4 \cdot p^5$

   a) p^{20} b) $(p^2)^{20}$ c) p^9 d) $2p^9$

2. $(-r^3)^4$

   a) $-r^7$ b) r^{12} c) $\dfrac{1}{r^7}$ d) $-r^{12}$

3. $\left(\dfrac{-2a}{b^2c^3}\right)^3$

   a) $\dfrac{-6a}{b^2c^3}$ b) $\dfrac{-8a^3}{b^2c^3}$ c) $\dfrac{-8a^3}{b^6c^9}$ d) $8a^3b^6c^9$

4. $\left(\dfrac{8w^2z}{32wz^4}\right)^3$

   a) $\dfrac{w}{12z^7}$ b) $\dfrac{w^6}{4z^{12}}$ c) $\dfrac{w^3}{64z^9}$ d) $\dfrac{w^3}{64z^{12}}$

5. 5^{-2}

   a) -25 b) $\dfrac{1}{25}$ c) -10 d) $-\dfrac{1}{10}$

6. $\dfrac{1}{t^{-3}}$

   a) $\dfrac{3}{t}$ b) $3t$ c) $-\dfrac{1}{t^3}$ d) t^3

7. $(7u^3v^{-1})(3u^{-2}v^5)$

   a) $21uv^4$ b) $\dfrac{10}{u^6v^5}$ c) $\dfrac{7v^4}{3u^2}$ d) $\dfrac{21}{uv^4}$

8. $\left(\dfrac{20y^2z^{-3}}{40yz^{-2}}\right)^{-1}$

   a) $2yz$ b) $\dfrac{2z}{y}$ c) $\dfrac{z}{2y}$ d) $\dfrac{2y}{z}$

Practice Test 4B

Write each of the following without the given metric prefix.

9. 41.7 kilowatts
   a) 0.0417 watts
   b) 41,700 watts
   c) 4170 watts
   d) 41,700,000 watts

10. 5 micrograms
    a) 0.0000005 gram
    b) 0.000005 gram
    c) 5,000,000 grams
    d) 50,000,000 grams

Perform each indicated operation and write your answer in scientific notation.

11. $\dfrac{2.0 \times 10^7}{2.5 \times 10^3}$
    a) 8.0×10^{10}
    b) 8.0×10^3
    c) 0.8×10^4
    d) 8.0×10^5

12. $(1.25 \times 10^2)(4.0 \times 10^{-4})$
    a) 5.0×10^{-2}
    b) 50×10^{-2}
    c) 0.5×10^{-2}
    d) 5.0×10^6

Convert each number to scientific notation, then calculate, and write your answer in scientific notation.

13. (7800)(0.001)
    a) 7.8×10^1
    b) 7.8×10^0
    c) 78×10^{-1}
    d) 7.8×10^6

14. $\dfrac{3,000,000}{0.15}$
    a) 2.0×10^5
    b) 2.0×10^6
    c) 2.0×10^7
    d) 2.0×10^8

Perform the indicated operations.

15. $(3t^2 + 7t - 4) + (-t^2 + 2t - 7)$
    a) $t^2 + 9t - 3$
    b) $9t + 8$
    c) $2t^2 + 9t + 11$
    d) $2t^2 + 9t - 11$

16. $\left(\dfrac{1}{2}p^2 - 6\right) + \left(\dfrac{1}{8}p^2 - 5p + 9\right)$
    a) $\dfrac{5}{8}p^2 - 5p + 3$
    b) $\dfrac{5}{8}p^2 - 11p + 9$
    c) $\dfrac{5}{8}p^2 - 5p - 3$
    d) $\dfrac{3}{8}p^2 - 11p + 9$

17. $(m^2 - 9) - (m - 3)$
    a) $m + 6$
    b) $m^2 - m - 6$
    c) $m^2 - m + 6$
    d) $m^2 - m + 12$

18. Subtract $(4c^2 - 7)$ from $(c^2 - c + 1)$.
    a) $-3c^2 + c + 8$
    b) $-3c^2 - c + 8$
    c) $3c^2 - c - 8$
    d) $3c^2 - c + 8$

19. $(6r + 1)(5r - 6)$
    a) $11r^2 - 5$
    b) $30r^2 - 31r + 6$
    c) $30r^2 - 31r - 6$
    d) $30r^2 - 41r - 6$

Practice Test 4B

20. $-7s(2s^2 + 3s - 7)$

    a) $2s^2 - 105 - 7$ b) $2s^2 - 4s - 7$ c) $-14s^3 - 21s^2 + 49s$ d) $-14s^3 - 21s^2 - 49s$

21. $\dfrac{6h^3 - 9h^2 + 18h}{3h}$

    a) $3h^2 - 6h + 15$ b) $2h^3 - 3h^2 + 6h$ c) $2h^2 - 3h + 6h$ d) $2h^2 - 3h + 6$

22. $\dfrac{12n^2 + 11n - 5}{3n - 1}$

    a) $4n^2 + 11n + 5$ b) $4n + 5$ c) $4n - 5$ d) $12n^2 + 8n - 4$

23. If a country is in debt by 3.3 trillion dollars and has a population of 300 million people, use scientific notation to determine how much each person owes.
    a) $11,000   b) $110,000   c) $11,000,000   d) $1100

24. Find the perimeter of a regular hexagon (6-sided figure) with each side measuring $(3x - 4)$ cm.

    a) $(3x + 2)$ cm   b) $(18x - 24)$ cm   c) $(3x - 4)$ cm   d) $(9x^2 - 24x + 16)$ cm

25. Find the area of a trapezoid with the longer base measuring $(5n + 1)$ in., the shorter base measuring $(3n - 1)$ in., and a height of $3n$ in.

    a) $24n^2 + 6n$ in.^{2} b) $24n^2$ in.^{2} c) $45n^3 - 6n^2 - 3n$ in.^{2} d) $12n^2$ in.^{2}

Notes:

Name:
Instructor:

Date:
Section:

5.1 Factoring a Monomial from a Polynomial

> **Objectives**
> **1** Identify factors.
> **2** Determine the greatest common factor of two or more numbers.
> **3** Determine the greatest common factor of two or more terms.
> **4** Factor a monomial from a polynomial.
>
> **Key Vocabulary**
> factor an expression, factors, prime number, composite number, prime factorization, GCF (greatest common factor) of two or more numbers, GCF of two or more terms

1 Identify factors.

Example 1 List the factors of the following:

a) 16   b) $9x^2$ c) $(2x+3)(5x-4)$

2 Determine the greatest common factor of two or more numbers.

Example 2 Write each as a product of prime factors.

a) 64   b) 96   c) 250

Example 3 Determine the GCF of the following pairs.

a) 24, 100   b) 39, 91   c) 32, 128

3 Determine the greatest common factor of two or more terms.

Example 4 Determine the GCF of each group of terms.

a) $3b^2, 15b^4, 21b^5$ b) $2p, 3r, 5t$ c) $9(x-1), 11(x-1)$

4 Factor a monomial from a polynomial.

Example 5 Factor out the GCF from the expressions.

a) $6n^2 - 36n$ b) $a^2 + 42a$ c) $7s^3 + 14s^2 - 35$

d) $19m^2 - m$ e) $17(r-3) + 2r(r-3)$ f) $75c^2 - 15c + 60$

Answers: 1. a) 1, 2, 4, 8, 16 b) 1, 3, 9, x, x^2, $3x$, $3x^2$, $9x$, $9x^2$ c) $(2x+3)$, $(5x-4)$ 2. a) 2^6 b) $2^5 \cdot 3$ c) $2 \cdot 5^3$ 3. a) 4 b) 13 c) 32
4. a) $3b^2$ b) 1 c) $x-1$ 5. a) $6n(n-6)$ b) $a(a+42)$ c) $7(s^3+2s^2-5)$ d) $m(19m-1)$ e) $(r-3)(17+2r)$ f) $15(5c^2-c+4)$

Practice Set 5.1

Write each number as a product of prime numbers.

1. 168

2. 270

3. 500

Determine the greatest common factor for each group of items.

4. 24, 72, 84

5. 6, 11, 17

6. uv, uv^2, $(uv)^2$

7. $w(w-2)$, $w-2$

8. $5n(7n+2)$, $-3n(7n+2)$

Factor the greatest common factor from each term in the expression and rewrite in factored form.

9. $7p - 35$

10. $42a^2bc^3 + 42a^3bc$

11. $9p^3 - 27p^2 - 36p$

12. $k(k-6) + 8(k-6)$

13. $7h(3h-1) + 3h - 1$

14. $15m^2n - 5mn^2 + 30m$

15. $4r^2(5r-1) - 7r(5r-1)$

1. _____
2. _____
3. _____
4. _____
5. _____
6. _____
7. _____
8. _____
9. _____
10. _____
11. _____
12. _____
13. _____
14. _____
15. _____

Name:
Instructor:
Date:
Section:

5.2 Factoring by Grouping

Objectives
1. Factor a polynomial containing four terms by grouping.

Key Vocabulary
factoring by grouping

1 Factor a polynomial containing four terms by grouping.

Example 1 Use the following steps to factor a four-term polynomial using grouping:

1) Determine whether there are any two factors common to all four terms. If so, factor the greatest common factor (GCF) from each term.

2) If necessary, arrange the four terms so that the first two items have a common factor and the last two have a common factor.

3) Use the distributive property to factor each group of two terms.

4) Factor out the GCF from the results in step 3.

a) $3x^2 - 5x + 12x - 20$

b) $2a^2 - 12ab - 7ab + 42b^2$

c) $y^2 - 3yz + 9yz - 27z^2$

d) $20p^2 - 35pr + 8pr - 14r^2$

e) $3x^2 + 12x + x + 4$

f) $14y^2 - 2yz - 7y + z$

Answers: 1a) $x(3x-5) + 4(3x-5)$; $(3x-5)(x+4)$ b) $2a(a-6b) - 7b(a-6b)$; $(a-6b)(2a-7b)$
c) $y(y-3z) + 9z(y-3z)$; $(y-3z)(y+9z)$ d) $5p(4p-7r) + 2r(4p-7r)$; $(4p-7r)(5p+2r)$
e) $3x(x+4) + 1(x+4)$; $(3x+1)(x+4)$ f) $2y(7y-z) - 1(7y-z)$; $(2y-1)(7y-z)$

Name:
Instructor:

Date:
Section:

Practice Set 5.2

Factor by grouping.

1. $p^2 + 5p + 12p + 60$

    1. _____

2. $k^2 - 8k + 9k - 72$

    2. _____

3. $h^2 + 3h - 15h - 45$

    3. _____

4. $g^2 - 7g - 8g + 56$

    4. _____

5. $f^2 - 10f - 12f + 120$

    5. _____

6. $v^2 + 8v - 18v - 144$

    6. _____

Rearrange the terms so that the first two terms have a common factor and the last two terms have a common factor. Next, factor by grouping. There may be more than one way to arrange the factors; however, the answers should be equivalent.

7. $9t - 2s + 18 - st$

    7. _____

8. $7n - 4m - mn + 28$

    8. _____

9. $bc + 40 - 5b - 8c$

    9. _____

10. $11d + 6c + 66 + cd$

    10. _____

Name:
Instructor:

Date:
Section:

5.3 Factoring Trinomials of the Form $ax^2 + bx + c$, $a = 1$

Objectives

1. Factor trinomials of the form $ax^2 + bx + c$, where $a = 1$.
2. Remove the greatest common factor from a trinomial.

Key Vocabulary
factoring by trial and error (reverse FOIL method), prime polynomial

1 Factor trinomials of the form $ax^2 + bx + c$, where $a = 1$.

Example 1 Consider the trinomial $x^2 + 5x - 24$.

a) List the factors of 24.

b) List the possible factors of the trinomial.

c) Determine the signs that will appear in the binomial factors.

d) Use trial and error to factor the trinomial.

Example 2 Consider the trinomial $x^2 - 9x + 20$.

a) List the factors of 20.

b) List the possible factors of the trinomial.

c) Determine the signs that will appear in the binomial factors.

d) Use trial and error to factor the trinomial.

2 Remove the greatest common factor from a trinomial.

Example 3 Consider the trinomial $3k^2 - 30k + 27$.

a) Factor out the factors common to all terms.

b) List the possible factors of the remaining trinomial.

c) Determine the signs that will appear in the binomial factors.

d) Use trial and error to factor the trinomial.

Example 4 Factor each polynomial. If the polynomial is prime, so state.

a) $c^2 - 3c - 70$ b) $z^2 - 13z + 9$ c) $p^2 - 19p + 48$

d) $-m^2 - 13m - 42$ e) $t^2 - st - 20s^2$

Answers: 1a) 1, 2, 3, 4, 6, 8, 12, 24 b, c) $(x+1)(x-24)$; $(x+2)(x-12)$; $(x+3)(x-8)$; $(x+4)(x-6)$; $(x-1)(x+24)$; $(x-2)(x+12)$; $(x-3)(x+8)$; $(x-4)(x+6)$ d) $(x-3)(x+8)$ 2a) 1, 2, 4, 5, 10, 20 b, c) $(x+1)(x+20)$; $(x+2)(x+10)$; $(x+4)(x+5)$; $(x-1)(x-20)$; $(x-2)(x-10)$; $(x-4)(x-5)$ d) $(x-4)(x-5)$ 3a) $3(k^2-10k+9)$ b, c) $3(k-1)(k-9)$; $3(k-3)(k-3)$; $3(k+1)(k+9)$; $3(k+3)(k+3)$ d) $3(k-1)(k-9)$ 4a) $(c+7)(c-10)$ b) prime c) $(p-16)(p-3)$ d) $-1(m+6)(m+7)$ e) $(t-5s)(t+4s)$

Copyright © 2015 by Pearson Education, Inc.   Angel/Runde, *Elementary Algebra for College Students*, 9e

Name:
Instructor:
Date:
Section:

Practice Set 5.3

Factor each polynomial. If the polynomial is prime, so state.

1. $a^2 - 4ab - 12b^2$ 1. _____

2. $u^2 - 18u + 81$ 2. _____

3. $t^2 - 5t - 50$ 3. _____

4. $g^2 + 2gh - 15h^2$ 4. _____

5. $s^2 + 11s - 26$ 5. _____

6. $y^2 - y - 30$ 6. _____

7. $k^2 - 5k - 6$ 7. _____

8. $p^2 - 4p + 12$ 8. _____

9. $3c^2 + 30c + 63$ 9. _____

10. $b^2 - 77 + 4b$ 10. _____

11. $-f^2 + 15f - 54$ 11. _____

12. $m^2 + 17m + 39$ 12. _____

13. $-2n^3 - 34n^2 - 140n$ 13. _____

14. $h^2 - 9h + 81$ 14. _____

15. $r^2 - 23r + 102$ 15. _____

16. $v^2 + 27vw + 72w^2$ 16. _____

17. $7w^5 + 7w^4 - 392w^3$ 17. _____

18. $q^2 + 9q - 70$ 18. _____

19. $y^2 - 12y - 160$ 19. _____

20. $36 - 13z + z^2$ 20. _____

Angel/Runde, *Elementary Algebra for College Students*, 9e

Name:
Instructor:
Date:
Section:

5.4 Factoring Trinomials of the Form $ax^2 + bx + c$, $a \neq 1$

Objectives
1. Factor trinomials of the form $ax^2 + bx + c$, $a \neq 1$, by trial and error.
2. Factor trinomials of the form $ax^2 + bx + c$, $a \neq 1$, by grouping.

1 Factor trinomials of the form $ax^2 + bx + c$, $a \neq 1$, by trial and error.

Example 1 Consider the trinomial $2t^2 + t - 6$.

a) List all pairs of factors of the coefficient of the squared term.

b) List all pairs of factors of the constant term.

c) Determine the signs that will appear in the binomial factors.

d) Try various combinations of the factor pairs until the correct middle term is found.

2 Factor trinomials of the form $ax^2 + bx + c$, $a \neq 1$, by grouping.

Example 2 Consider the trinomial $2k^2 - 41kz + 20z^2$.

a) Find two numbers whose product is equal to the product of a times c, and whose sum is equal to b.

b) Rewrite the middle term, bx, as the sum or difference of two terms using the numbers found in step a).

c) Factor the expanded polynomial by grouping.

Example 3 Consider the trinomial $8w^3 - 44w^2 - 160w$.

a) Determine whether there is a factor common to all three terms; if so, factor out the GCF.

b) Factor using trial and error.

c) Factor using grouping.

d) Check your answer by using FOIL and then distribute the common factor.

Answers: 1a) 1, 2   b) 1, –6, 2, –3, 3, –2, 6, –1   c,d) $(2t+1)(t-6)$; $(2t+2)(t-3)$; $(2t+3)(t-2)$; $(2t+6)(t-1)$; $(2t-1)(t+6)$; $(2t-2)(t+3)$; $(2t-3)(t+2)$; $(2t-6)(t+1)$; Answer: $2t^2 + t - 6 = (2t-3)(t+2)$ 2a) product of $a \cdot c = 40$; $b = -41$
b) $-41 = -1 - 40$ c) $2k^2 - 1kz - 40kz + 20z^2 = k(2k-z) - 20z(2k-z) = (k-20z)(2k-z)$
3a) $4w(2w^2 - 11w - 40)$ b) $4w(2w+5)(w-8)$ c) $a \cdot c = -80$; $b = -11$

Factors of $a \cdot c$	Sum of factors
$-80 \cdot 1$	$-80 + 1 = -79$
$-40 \cdot 2$	$-40 + 2 = -38$
$-20 \cdot 4$	$-20 + 4 = -16$
$-16 \cdot 5$	$-16 + 5 = -11$*
$-10 \cdot 8$	$-10 + 8 = -2$

Use factors -16 and 5: $2w^2 - 11w - 40 = 2w^2 - 16w + 5w - 40 = 2w(w-8) + 5(w-8) = (2w+5)(w-8)$;

Answer: $8w^3 - 44w^2 - 160w = 4w(2w+5)(w-8)$ d) $4w(2w+5)(w-8) = 4w(2w^2 - 16w + 5w - 40) = 4w(2w^2 - 11w - 40) = 8w^3 - 44w^2 - 160w$

Name:
Instructor:

Date:
Section:

Practice Set 5.4

Factor completely. If the polynomial is prime, so state.

1. $20v^2 - v - 1$
2. $18m^2 + 27m + 4$

3. $8r^2 - 13r + 5$
4. $6c^2 - c - 15$

5. $10k^2 - 29k + 10$
6. $20n^2 + 37n + 15$

7. $7t^2 + 47t - 14$
8. $6y^2 + 5y + 1$

9. $11b^2 - 54b - 5$
10. $-24h^4 - 88h^3 - 32h^2$

11. $4r^2 - 20r + 25$
12. $18v^2 + 55vw - 28w^2$

13. $12y^2 - 145yz + 12z^2$
14. $15f^2 - 32fg + 16g^2$

15. $22p^2 + 51p - 10$
16. $33s^2 + 34st - 35t^2$

17. $24d^2 + 41d + 12$
18. $12g^2 + 19g + 5$

19. $18u^2 - 39u + 20$
20. $15 - 41x + 14x^2$

1. _____
2. _____
3. _____
4. _____
5. _____
6. _____
7. _____
8. _____
9. _____
10. _____
11. _____
12. _____
13. _____
14. _____
15. _____
16. _____
17. _____
18. _____
19. _____
20. _____

Name:
Instructor:
Date:
Section:

5.5 Special Factoring Formulas and a General Review of Factoring

Objectives
1. Factor the difference of two squares.
2. Factor the sum and difference of two cubes.
3. Learn the general procedure for factoring a polynomial.

Key Vocabulary
difference of two squares, sum of two cubes, difference of two cubes

1 Factor the difference of two squares.

Example 1 Write the difference of two squares formula.

Example 2 Factor using the difference of two squares formula.

a) $r^2 - 64$ b) $1 - 81c^2$ c) $121w^2 - 4z^2$

2 Factor the sum and difference of two cubes.

Example 3

a) Write the sum of two cubes formula.

b) Write the difference of two cubes formula.

Example 4 Factor using the sum and difference of two cubes formulas.

a) $8t^3 - 1$ b) $v^3 + 125$ c) $27a^3 - 216b^3$

3 Learn the general procedure for factoring a polynomial.

Example 5 Factor each completely using the general procedure for factoring a polynomial.

a) $15c - 33$ b) $2x^2 + 2xy - 8xz - 8yz$ c) $m^2n + 4mn - 32n$

d) $3s^3 - 12st^2$ e) $250x^4 + 128xy^3$

Answers: 1. $a^2 - b^2 = (a+b)(a-b)$ 2a) $(r+8)(r-8)$ b) $(1+9c)(1-9c)$ c) $(11w+2z)(11w-2z)$
3a) $a^3 + b^3 = (a+b)(a^2 - ab + b^2)$ b) $a^3 - b^3 = (a-b)(a^2 + ab + b^2)$ 4a) $(2t-1)(4t^2 + 2t + 1)$
b) $(v+5)(v^2 - 5v + 25)$ c) $27(a-2b)(a^2 + 2ab + 4b^2)$ 5a) $3(5c-11)$ b) $2(x+y)(x-4z)$ c) $n(m+8)(m-4)$
d) $3s(s+2t)(s-2t)$ e) $2x(5x+4y)(25x^2 - 20xy + 16y^2)$

Name:
Instructor:

Date:
Section:

Practice Set 5.5

Factor completely. If the polynomial is prime, so state.

1. $35s^3 - 28s^2$

2. $1 - 125t^3$

3. $m^3 + 64n^3$

4. $400a^2b^2 - c^2$

5. $16v^2 + 25w^2$

6. $3x^2 + 27x + 60$

7. $2k + 2h + jk + jh$

8. $63r^2 + 42pr + 7p^2$

9. $121z^2 - 22z + 1$

10. $3t^3 - 12tv^2$

11. $p^3 + 2p^2 + 3p + 6$

12. $4mn^2 + 7mn - 15m$

13. $8u^3 - 36u^2 - 20u$

14. $-54f^3 + 250g^3$

15. $12m^3n - 75mn^3$

16. $a^2 + 2a - 9a - 18$

17. $mx + 3ax + my + 3ay$

18. $-12p^2 + 5p + 2$

19. $6a^2 - 54b^2$

20. $84 - 7x - 7x^2$

1. _____
2. _____
3. _____
4. _____
5. _____
6. _____
7. _____
8. _____
9. _____
10. _____
11. _____
12. _____
13. _____
14. _____
15. _____
16. _____
17. _____
18. _____
19. _____
20. _____

Name:
Instructor:

Date:
Section:

5.6 Solving Quadratic Equations Using Factoring

Objectives
1. Recognize quadratic equations.
2. Solve quadratic equations using factoring.

Key Vocabulary
quadratic equation, standard form of a quadratic equation, zero-factor property

1 Recognize quadratic equations.

Example 1 Answer yes or no.

a) Is $4p - 3 = 0$ a quadratic equation?
b) Is $6z^2 - 2z = 9$ a quadratic equation?
c) Is $8r^2 - 8 = 0$ a quadratic equation?
d) Is $9m + 10 = m^2$ a quadratic equation?
e) Is $3x^2 + 2x + 1 = 3x^2$ a quadratic equation?

Example 2 Solve the following equations.

a) $(7t - 4)(2t + 5) = 0$ b) $6m(9m - 1) = 0$ c) $(8c + 3)(8c - 3) = 0$

2 Solve quadratic equations using factoring.

Example 3 Consider the equation $21z^2 = 17z - 2$.

a) Write the equation in standard form with the squared term having a positive coefficient. This will result in one side of the equation being 0.

b) Factor the side of the equation that is not 0.

c) Set each factor containing a variable equal to 0 and solve each equation.

d) Check each solution in the original equation.

Answers: 1a) no b) yes c) yes d) yes e) no   2a) $\frac{4}{7}, -\frac{5}{2}$ b) $0, \frac{1}{9}$ c) $-\frac{3}{8}, \frac{3}{8}$

3a) $21z^2 - 17z + 2 = 0$ b) $(3z - 2)(7z - 1) = 0$ c) $z = \frac{2}{3}$ or $z = \frac{1}{7}$ d) Both solutions check.

Name:
Instructor:

Date:
Section:

Practice Set 5.6

Solve.

1. $x(2x-13) = -15$
2. $3m^2 + 6m = 0$

3. $2(y-4) = -y^2$
4. $p(p+10) = 2(p-8)$

5. $2c(c-3) = 3-c$
6. $4n^2 - 5 = n$

7. $6u^2 - 11u = -5$
8. $(g-4)^2 = 1$

9. $2r(3r+10) = -6$
10. $a^2 + 7a = a - 9$

11. $2k^2 + 5k = 3$
12. $9v^2 = 4$

13. $(5+d)(7-d) = -13$
14. $3t^2 - 48 = 0$

15. $s^2 = 1$
16. $35f^2 + 13f - 4 = 0$

17. $2w - w^2 + 24 = 0$
18. $b^2 - 121 = 0$

19. $h^2 - 8h + 7 = 0$
20. $-3(z-6) + 2 = z^2 + 2$

1. _____
2. _____
3. _____
4. _____
5. _____
6. _____
7. _____
8. _____
9. _____
10. _____
11. _____
12. _____
13. _____
14. _____
15. _____
16. _____
17. _____
18. _____
19. _____
20. _____

Name:
Instructor:

Date:
Section:

5.7 Applications of Quadratic Equations

> **Objectives**
> **1** Solve applications by factoring quadratic equations.
> **2** Learn the Pythagorean Theorem.
>
> **Key Vocabulary**
> right triangle, legs of a right triangle, hypotenuse, Pythagorean Theorem

1 Solve applications by factoring quadratic equations.

Express each problem as an equation, then solve.

Example 1

The product of two consecutive odd positive integers is 255. Find the integers.

Example 2

A realtor suggests to a client that his house would sell faster if new carpeting was installed. While measuring, the client discovered that the length of the master bedroom was twice as long as the width. If the area is 288 square feet, determine the dimensions of the room.

2 Learn the Pythagorean Theorem.

Example 3

A 30-foot ladder is leaning up against the side of a building. The base of the ladder is 18 feet from the building. How far up the building does the ladder reach?

Example 4

The diagonal of a rectangular table is 5 feet long. How long are the table's sides if the length is one foot longer than the width?

Answers: 1) 15, 17 2) 12 ft by 24 ft 3) 24 ft 4) 3 ft by 4 ft

Name:
Instructor:

Date:
Section:

Practice Set 5.7

1. One negative number is 5 less than another. Their product is 84. Find the numbers.

    1. _____

2. If 4 is added to the square of a number, the result is 5 less than 10 times that number. What is the number?

    2. _____

3. The area of a square is numerically equal to its perimeter. How long is a side?

    3. _____

4. A triangle has an area of 10 cm² and its height is 3 cm less than twice the length of its base. Find the base and height of the triangle.

    4. _____

5. Two positive numbers have a difference of 12 and a product of 45. What are the numbers?

    5. _____

6. A rectangle with a base of x in. and a height that is 2 in. less than the base, is inscribed in a circle with a diameter of 10 in. Find the base and height of the rectangle.

    6. _____

7. A 20-foot tall tower is secured by 3 guy wires fastened at the top and to anchors 15 ft from the base of the tower. How long is each wire?

    7. _____

8. An object was thrown with an initial velocity of 128 ft/sec. Use the formula $h = vt - 16t^2$, where h is the height in feet and t is time in seconds, to find the number of seconds it will take for the object to return to the ground.

    8. _____

9. The three side lengths of a right triangle are three consecutive integers. Find the lengths of the three sides in inches. (Hint: The hypotenuse must be the longest side.)

    9. _____

10. Find the width of a rectangular solid whose length is 10 cm and its height is 1 cm longer than twice its width. It has a volume of 210 cm³.

    10. _____

Chapter 5 Vocabulary Reference Sheet

Term	Definition	Example
factors	If $a \cdot b = c$, then a and b are factors of c.	In $8x$, 8 and x are factors of $8x$.
factor an expression	To write the expression as a product of its factors.	$3x^3 - 27xy^2$ written in factored form is $3x(x-3y)(x+3y)$.
prime number	An integer greater than 1 that has exactly two factors, itself and 1	3, 5, 7, 11, and 13 are prime numbers.
composite number	A positive integer (other than 1) that is not prime	2, 4, 6, 9, and 16 are composite numbers.
unit	The number 1, which is neither prime nor composite	1
GCF of two or more numbers	The greatest number that divides evenly into all of the numbers	The GCF of 6, 12, and 15 is 3. The GCF of 7 and 12 is 1.
GCF of two or more terms	Take each factor the largest number of times it appears in all of the terms.	The GCF of $2p^2$, $6p$, and $10p^3$ is $2p$.
factor a monomial from a polynomial	Determine the GCF of all the terms in the polynomial. Use the distributive property to factor out the GCF.	$9u^2 - 6u$ factors as $3u(3u-2)$.
factoring by grouping	Factoring a polynomial containing four terms by removing common factors from pairs of terms.	$b^2 - 3b + 7b - 21 = b(b-3) + 7(b-3)$ $= (b-3)(b+7)$
prime polynomial	A polynomial that cannot be factored using only integer coefficients	$5x^2 + 6x + 7$ is a prime polynomial.
factor a trinomial of the form $ax^2 + bx + c$	Factor out the GCF. If $a = 1$, use the trial-and-error method. If $a > 1$, use factor by grouping or trial and error.	$40v^2 - 14v - 6$ $= 2(20v^2 - 7v - 3)$ $= 2(5v-3)(4v+1)$
difference of two squares	$a^2 - b^2 = (a+b)(a-b)$	$9 - m^2 = (3+m)(3-m)$.
sum of two cubes	$a^3 + b^3 = (a+b)(a^2 - ab + b^2)$	$64n^3 + 125m^3$ $= (4n+5m)(16n^2 - 20mn + 25m^2)$
difference of two cubes	$a^3 - b^3 = (a-b)(a^2 + ab + b^2)$	$8p^3 - 27 = (2p-3)(4p^2 + 6p + 9)$
quadratic equation	An equation of the form $ax^2 + bx + c = 0$, where a, b, and c are real numbers, $a \neq 0$.	$x^2 + 4x - 12 = 0$, $2x^2 - 5x = 0$, and $3x^2 - 2 = 0$ are quadratic equations in **standard form**.
zero-factor property	If $ab = 0$, then $a = 0$ or $b = 0$.	If $z(z-3) = 0$, then $z = 0$ or $z - 3 = 0$.
solve a quadratic equation	Set all terms in descending order equal to 0. Factor the non-zero side of the equation. Set each factor equal to zero (apply the zero-factor property) and solve.	Solve $t^2 - 3t = 10$: $t^2 - 3t - 10 = 0$ $(t-5)(t+2) = 0$ $t - 5 = 0$ or $t + 2 = 0$ $t = 5$ or $t = -2$

Chapter 5 Vocabulary

right triangle	A triangle that contains a right, or 90°, angle. The **legs** form the right angle, and the side across from the right angle is the **hypotenuse**.	leg = b, hypotenuse = c, leg = a
Pythagorean Theorem	If a and b represent the legs of a right triangle, and c represents the hypotenuse, then $a^2 + b^2 = c^2$.	In the triangle below, $3^2 + 4^2 = 5^2$. (sides 3, 4, 5)

Name:
Instructor:

Date:
Section:

Chapter 5 Practice Test A

Find the greatest common factor (GCF) for each set of terms.

1. $4a,\ 8a^2,\ 32a^3$

2. $7p,\ 9n,\ 10$

Factor each expression completely. If an expression is prime, so state.

3. $72y^3 - 48y^2$

4. $13rs - 91r^2s^2$

5. $2t^2 + 2tu - 3t - 3u$

6. $6w^2 - 35 + 42w - 5w$

7. $h^2 + 22h + 72$

8. $m^2 - 10m - 56$

9. $v^2 + 9v + 36$

10. $1 + 125b^3$

11. $6c^2 + c - 35$

12. $25k^2 + 30k + 9$

13. $196n^2 - 1$

14. $25x^2 + y^2$

15. $4y^3 - 32z^3$

1. _____

2. _____

3. _____

4. _____

5. _____

6. _____

7. _____

8. _____

9. _____

10. _____

11. _____

12. _____

13. _____

14. _____

15. _____

Practice Test 5A

Solve.

16. $4p^2 = 9$

16. _____

17. $6c^2 - 7c - 3 = 0$

17. _____

18. $z(z - 2) = 63$

18. _____

19. $2(x + 1)^2 = 18$

19. _____

Solve.

20. Samuel is 12 years older than his brother James. The product of their ages is 133. How old is each brother?

20. _____
_____

21. The product of two positive consecutive even integers is 224. What are the numbers?

21. _____

22. A 30-foot ladder is leaning up against a tree. It meets the tree 24 ft up the tree. How far from the tree is the base of the ladder?

22. _____

Name:
Instructor:

Date:
Section:

Chapter 5 Practice Test B

1. Determine the GCF of $32m^2$, $16m$, and $24m^3$.
   a) $4m$ b) 8 c) $8m$ d) none of these

2. Determine the GCF of $9c$, $90c^4$, and 42.
   a) $9c$ b) 3 c) $3c$ d) none of these

3. Factor completely: $12z^2 - 24z$
   a) $12z(z-2)$ b) $z(12z-2)$ c) $2x(6z-12)$ d) none of these

4. Factor completely: $m^3 - 9m$
   a) $m(m-3)(m-3)$ b) $m(m-3)(m+3)$ c) $2x(6z-12)$ d) none of these

5. Find one of the factors of $3m^2 - 5m - 22$.
   a) $(3m+2)$ b) $(3m-11)$ c) $(m-11)$ d) none of these

6. Find one of the factors of $t^2 + 4t - 32$.
   a) $(t+8)$ b) $(t+4)$ c) $(t-8)$ d) none of these

7. Find one of the factors of $7n^2 + 31n + 12$.
   a) $(7n+4)$ b) $(n+12)$ c) $(n+4)$ d) none of these

8. Find one of the factors $5b^2 - 22b + 8$.
   a) $(5b+2)$ b) $(b+4)$ c) $(5b+4)$ d) none of these

9. Find one of the factors of $w^3 - 1000$.
   a) $(w^2 - 10w + 100)$ b) $(w^2 + 10w + 100)$ c) $(w^2 + 10w - 100)$ d) none of these

10. Find one of the factors of $6v^3 - 5v^2 + v$.
    a) $(2v^2 - 1)$ b) $(3v^2 + 1)$ c) $(2v - 1)$ d) none of these

11. Find one of the factors of $5h^2 + 6h + 15$.
    a) $(5h+3)$ b) $(h+5)$ c) $(5h+5)$ d) none of these

Practice Test 5B

12. Find one of the factors of $k^2 - 4k + 4$.
    a) $(k+4)$ b) $(k+2)$ c) $(k-2)$ d) none of these

13. Find one of the factors of $12z^2 - 13z + 3$.
    a) $(4z+3)$ b) $(4z-1)$ c) $(4z-3)$ d) none of these

14. Find one of the factors of $18r^2 - 19r + 5$
    a) $(2r-1)$ b) $(2r+1)$ c) $(2r+5)$ d) none of these

15. Solve $g^2 - 7g = -12$.
    a) 6, 2   b) 4, 3   c) 12, 1   d) −12, 1

16. Solve $12p^2 = 4p + 1$.
    a) 6, 2   b) $-\dfrac{1}{6}, \dfrac{1}{2}$ c) $\dfrac{1}{6}, -\dfrac{1}{2}$ d) −6, −2

17. Solve $-4u(u-4) = 16$.
    a) 4, −4   b) 0, 4   c) 2   d) −2, 2

18. Solve $9y^2 = 16$.
    a) 9, 16   b) $\dfrac{1}{9}, \dfrac{1}{16}$ c) $\dfrac{3}{4}, -\dfrac{3}{4}$ d) $\dfrac{4}{3}, -\dfrac{4}{3}$

19. The product of two consecutive positive even integers is 224. What is the larger integer?
    a) 10   b) 12   c) 14   d) 16

20. A sail on a sailboat is a right triangle with a base that is 3 ft longer than its height. If the sail has a hypotenuse of 15 ft, how long is the base?
    a) 6 ft   b) 12 ft   c) 9 ft   d) 15 ft

Name: Date:
Instructor: Section:

6.1 Simplifying Rational Expressions

Objectives
1. Determine the values for which a rational expression is defined.
2. Understand the three signs of a fraction.
3. Simplify rational expressions.
4. Factor a negative 1 from a polynomial.

Key Vocabulary
rational expression, undefined, simplify/reduce to lowest terms

1 Determine the values for which a rational expression is defined.

Example 1 Determine the value or values of the variable for which the rational expression is defined.

a) $\dfrac{2}{x-5}$ b) $\dfrac{x+3}{4x+1}$ c) $\dfrac{2x}{x^2-4}$ d) $\dfrac{x+1}{x^2-x-6}$

2 Understand the three signs of a fraction.

Example 2 Write the rational expression $-\dfrac{x-2}{3}$ three other ways.

Example 3 Write the rational expression $-\dfrac{5x}{x+6}$ three other ways.

3 Simplify rational expressions.

Example 4 Simplify.

a) $\dfrac{9x^2 y}{6xy^2}$ b) $\dfrac{2x+6}{4x^2-36}$ c) $\dfrac{18a^2-6ab}{12a^2 b}$ d) $\dfrac{x^2-64}{x-8}$

4 Factor a negative 1 from a polynomial.

Example 5 Simplify.

a) $\dfrac{m-7}{7-m}$ b) $\dfrac{x^2-64}{8-x}$ c) $\dfrac{y^2-y-12}{8-2y}$ d) $\dfrac{x^2-9x+8}{1-x^2}$

Answers: 1a) $x \neq 5$ b) $x \neq -\dfrac{1}{4}$ c) $x \neq 2$ and $x \neq -2$ d) $x \neq 3$ and $x \neq -2$ 2. $\dfrac{-(x-2)}{3}, \dfrac{-x+2}{3}, \dfrac{2-x}{3}$ 3. $\dfrac{-5x}{x+6}, \dfrac{5x}{-(x+6)}, \dfrac{5x}{-x-6}$ 4a) $\dfrac{3x}{2y}$ b) $\dfrac{1}{2(x-3)}$ c) $\dfrac{3a-b}{2ab}$ d) $x+8$ 5a) -1 b) $-x-8$ c) $\dfrac{-y-3}{2}$ d) $\dfrac{8-x}{x+1}$

Copyright © 2015 by Pearson Education, Inc.

Name:
Instructor:

Date:
Section:

Practice Set 6.1

Determine the value or values of the variable that make each expression defined. Match the expression to the value or values on the right.

1. $\dfrac{x-3}{x+3}$

2. $\dfrac{x}{x^2-9}$

3. $\dfrac{x-3}{3x}$

4. $\dfrac{3x}{x-3}$

A. all real numbers except $x = 3$

B. all real numbers except $x = 0$

C. all real numbers except $x = -3$

D. all real numbers except $x = 3$, $x = -3$

E. all real numbers except $x = 9$

1. _____
2. _____
3. _____
4. _____

Simplify.

5. $\dfrac{18a^2b^4}{12a^3b^2}$

6. $\dfrac{5x-15y}{10x^2-30xy}$

7. $\dfrac{c-4}{3c^2-11c-4}$

8. $\dfrac{6y-24}{2y^2-32}$

9. $\dfrac{xy-7x+y-7}{x^2-1}$

10. $\dfrac{y^2+2y}{2y^2+8y+8}$

11. $\dfrac{5b-30}{5b^2-40b+60}$

12. $\dfrac{2z^2-6z-20}{z^2-8z+15}$

13. $\dfrac{1-2s}{2s^2+5s-3}$

14. $\dfrac{x^3-8}{x^2-4}$

15. $\dfrac{2x^2-6x+5x-15}{2x^2-x+6x-3}$

16. $\dfrac{3x^2-9x+12x-36}{9x+36}$

17. $\dfrac{2x^2-50}{(x-5)^2}$

18. $\dfrac{2r^2-24r+70}{14-2r}$

5. _____
6. _____
7. _____
8. _____
9. _____
10. _____
11. _____
12. _____
13. _____
14. _____
15. _____
16. _____
17. _____
18. _____

Name: Date:
Instructor: Section:

6.2 Multiplication and Division of Rational Expressions

Objectives
1. Multiply rational expressions.
2. Divide rational expressions.

1 Multiply rational expressions.

Example 1 Multiply.

a) $\dfrac{35}{6} \cdot \dfrac{18}{7}$

b) $\left(-\dfrac{20}{38}\right)\left(\dfrac{38}{20}\right)$

Example 2 Multiply.

a) $\dfrac{9x^2 y}{20xy^3} \cdot \dfrac{25}{6xy^2}$

b) $\dfrac{x^2 y - xy}{12x^2} \cdot \dfrac{9xy}{3x^2 - 3}$

c) $\dfrac{3x^2 + 2x - 8}{8x^2 - 16x} \cdot \dfrac{4x + 20}{3x^2 + 11x - 20}$

2 Divide rational expressions.

Example 3 Divide.

a) $\dfrac{64}{25} \div \dfrac{16}{20}$

b) $\left(-\dfrac{24}{30}\right) \div \left(\dfrac{24}{30}\right)$

Example 4 Divide.

a) $\dfrac{7m^2 n^3}{32n^2} \div \dfrac{14m}{8mn^4}$

b) $\dfrac{x^3 - 8}{2x^2 - 8} \div \dfrac{x^2 + 2x + 4}{10x^2}$

c) $\dfrac{18 - 9a}{18ab - 42b} \div \dfrac{3a^2 b - 12b}{3a^2 b^2 - ab^2 - 14b^2}$

_____
Answers: 1a) 15   b) -1 2a) $\dfrac{15}{8y^4}$ b) $\dfrac{y^2}{4(x+1)}$ c) $\dfrac{x+2}{2x(x-2)}$ 3a) $\dfrac{16}{5}$ b) -1 4a) $\dfrac{m^2 n^5}{8}$ b) $\dfrac{5x^2}{x+2}$ c) $-\dfrac{1}{2}$

Name:
Instructor:

Date:
Section:

Practice Set 6.2

Multiply.

1. $\dfrac{10x^2 y}{11x^3} \cdot \dfrac{33xy^2}{20y^4}$

2. $\dfrac{x-5}{5x^2} \cdot \dfrac{25x}{x^2-25}$

3. $\dfrac{a^2 - 2a + 1}{2a - 12} \cdot \dfrac{4a - 4}{(a-1)^2}$

4. $\dfrac{y^3 - 64}{y^2 + 4y + 16} \cdot \dfrac{6y}{3y^3 - 12y}$

Divide.

5. $\dfrac{18ab^2}{15a^3 b} \div \dfrac{12a^2 b}{10a^3 b^2}$

6. $\dfrac{s-3}{16s^3} \div \dfrac{2s^2 - 18}{8s^2}$

7. $\dfrac{9x^2 - 25}{27x^2 + 15x - 2} \div \dfrac{6x^2 - 7x - 5}{9x^2 - 4}$

8. $\dfrac{6x^2 - 54}{15x + 15} \div \dfrac{3x - 9}{5x^2 - 5}$

Perform the indicated operation.

9. $\dfrac{40x^2 y^3}{8xy^4} \cdot \dfrac{20xy}{16xy^2}$

10. $\dfrac{s - t}{s^2 - 11s + 10} \div \dfrac{s^2 - t^2}{s^2 - 12s + 20}$

11. $\dfrac{a^2 - 2a + 1}{2a - 12} \div \dfrac{4a - 4}{(a-1)^2}$

12. $\dfrac{x^2 - 7x - 8}{x^2 - 4x - 5} \div \dfrac{2x^2 - 15x - 8}{2x^2 - 7x - 4}$

13. $\dfrac{5x^2 - 10x}{x^2 - 49} \cdot \dfrac{9x^2 - 63x}{10x^3 - 40x}$

14. $\dfrac{25x^2 y^3}{4x - 12y} \cdot \dfrac{2x^2 - 18y^2}{50xy^4}$

15. $\dfrac{3x^2 + 5x + 2}{15x^2 - 7x - 2} \div \dfrac{3x^2 + 5x + 2}{3x^2 - 5x + 2}$

16. $\dfrac{4x + 20}{16x + 16} \cdot \dfrac{8x^2 + 8x}{2x^2 - 50}$

1. _____

2. _____

3. _____

4. _____

5. _____

6. _____

7. _____

8. _____

9. _____

10. _____

11. _____

12. _____

13. _____

14. _____

15. _____

16. _____

Name: Date:
Instructor: Section:

6.3 Addition and Subtraction of Rational Expressions with a Common Denominator and Finding the Least Common Denominator

Objectives
1. Add and subtract rational expressions with a common denominator.
2. Find the least common denominator.

Key Vocabulary
LCD (least common denominator)

1 Add and subtract rational expressions with a common denominator.

Example 1 Perform the indicated operation and simplify, if possible.

a) $\dfrac{7}{8} + \dfrac{9}{8}$

b) $\dfrac{9}{14} - \dfrac{11}{14}$

Example 2 Perform the indicated operation and simplify, if possible.

a) $\dfrac{6}{5x} + \dfrac{4}{5x}$

b) $\dfrac{3x^2}{x-2} - \dfrac{12}{x-2}$

c) $\dfrac{2x}{x^2-9x+20} - \dfrac{x+4}{x^2-9x+20}$

2 Find the least common denominator.

Example 3 Find the least common denominator of the following fractions.

a) $\dfrac{1}{4}, \dfrac{5}{8}$

b) $\dfrac{3}{5}, \dfrac{2}{3}$

c) $\dfrac{7}{10}, \dfrac{11}{5}, \dfrac{1}{6}$

Example 4 Find the least common denominator of the following rational expressions.

a) $\dfrac{1}{4x}, \dfrac{5}{8y}$

b) $\dfrac{x+1}{5x-10}, \dfrac{x}{10x}$

c) $\dfrac{x+3}{x^2+x-2}, \dfrac{x+2}{2x^2+x-3}$

d) $\dfrac{x+3}{x^2-1}, \dfrac{5x}{4x^2-4}$

Answers: 1a) 2 b) $-\dfrac{1}{7}$ 2a) $\dfrac{2}{x}$ b) $3x+6$ c) $\dfrac{1}{x-5}$ 3a) 8 b) 15 c) 30 4a) $8xy$ b) $10x(x-2)$ c) $(x+2)(x-1)(2x+3)$
d) $4(x-1)(x+1)$

Name:
Instructor:

Date:
Section:

Practice Set 6.3

Add or subtract.

1. $\dfrac{10}{3x} + \dfrac{5}{3x}$

2. $\dfrac{4}{12y} + \dfrac{y}{12y}$

3. $\dfrac{13a}{4b} - \dfrac{a}{4b}$

4. $\dfrac{x}{x^2-4} + \dfrac{2}{x^2-4}$

5. $\dfrac{rs}{r+2} - \dfrac{2rs}{r+2}$

6. $\dfrac{3x-2}{4x^2} - \dfrac{x+5}{4x^2}$

7. $\dfrac{3x-1}{x^2-1} + \dfrac{x-3}{x^2-1}$

8. $\dfrac{7t-5}{t-1} - \dfrac{8t-6}{t-1}$

9. $\dfrac{-x}{2x^2-8} + \dfrac{2}{2x^2-8}$

10. $\dfrac{4x^2-40}{2x^2-16x+30} - \dfrac{12x}{2x^2-16x+30}$

1. _____
2. _____
3. _____
4. _____
5. _____
6. _____
7. _____
8. _____
9. _____
10. _____

Find the least common denominator for the expressions on the left. Match the expressions to their LCD.

11. $\dfrac{x+5}{5}, \dfrac{x}{10}$ A. $5(x+5)$

12. $\dfrac{10}{x+5}, \dfrac{x}{5}$ B. $10x$

13. $\dfrac{3}{5x}, \dfrac{x}{10}$ C. 10

14. $\dfrac{10}{x^2-25}, \dfrac{x}{x^2}$ D. x^2-25

15. $\dfrac{10}{x+5}, \dfrac{x}{x-5}$ E. $5x(x+5)$

16. $\dfrac{3}{x+5}, \dfrac{x}{5x}$ F. $x^2(x^2-25)$

11. _____
12. _____
13. _____
14. _____
15. _____
16. _____

Challenge

17. Subtract:

$\dfrac{\Delta+8}{3\Delta+2} - \dfrac{\Delta+8}{3\Delta+2}$

18. Find the LCD:

$\dfrac{2}{5\Delta^2-20}, \dfrac{\Delta}{10\Delta^3-40\Delta}$

17. _____
18. _____

Name:
Instructor:

Date:
Section:

6.4 Addition and Subtraction of Rational Expressions

Objectives
1. Add and subtract rational expressions.

1 Add and subtract rational expressions.

Example 1 Add or subtract.

a) $\dfrac{1}{4} + \dfrac{5}{8}$
b) $\dfrac{3}{5} - \dfrac{2}{3}$
c) $\dfrac{7}{10} + \dfrac{11}{5} - \dfrac{1}{6}$

Example 2 Add or subtract.

a) $\dfrac{1}{4x} + \dfrac{5}{8y}$
b) $\dfrac{x+1}{5x-10} + \dfrac{x}{10x}$
c) $\dfrac{x+3}{x^2+x-2} - \dfrac{x+2}{2x^2+x-3}$

d) $\dfrac{x+3}{x^2-1} - \dfrac{5x}{4x^2-4}$
e) $\dfrac{x-5}{x-1} - \dfrac{2x-5}{1-x}$
f) $\dfrac{x-1}{9-x^2} + \dfrac{2}{7x^2-20x-3}$

Answers: 1a) $\dfrac{7}{8}$ b) $-\dfrac{1}{15}$ c) $\dfrac{41}{15}$ 2a) $\dfrac{5x+2y}{8xy}$ b) $\dfrac{3x}{10(x-2)}$ c) $\dfrac{x^2+5x+5}{(x+2)(2x+3)(x-1)}$ d) $\dfrac{12-x}{4(x-1)(x+1)}$ e) $\dfrac{3x-10}{x-1}$

f) $\dfrac{-7x^2+8x+7}{(7x+1)(x+3)(x-3)}$

Name:
Instructor:

Date:
Section:

Practice Set 6.4

Add or subtract.

1. $\dfrac{1}{a} + \dfrac{3}{b}$

2. $\dfrac{7x}{3} + \dfrac{2x}{3y}$

3. $\dfrac{2}{5x^2} - \dfrac{2x}{10}$

4. $\dfrac{9}{x-1} - \dfrac{7}{x+1}$

5. $\dfrac{x-y}{x+y} + \dfrac{2x}{x}$

6. $\dfrac{6}{x-2} - \dfrac{3x}{2-x}$

7. $\dfrac{3x}{x^2-1} + \dfrac{8}{(x-1)^2}$

8. $\dfrac{c}{2c-4} + \dfrac{8}{c-2}$

9. $\dfrac{9}{x^2-9x} - \dfrac{x}{9x-81}$

10. $\dfrac{x+6}{x^2-2x-3} - \dfrac{x+1}{x^2-x-6}$

11. $\dfrac{1}{6r^2+7r+2} + \dfrac{3}{2r^2-3r-2}$

12. $\dfrac{x-5}{x+5} + \dfrac{x+5}{x-5}$

13. $\dfrac{1}{9x^2-81} + \dfrac{2}{9x-27}$

14. $\dfrac{x}{14-5x-x^2} + \dfrac{4}{x^2-3x+2}$

15. $\dfrac{12}{x^2-4x} - \dfrac{3x}{4x-16}$

16. $\dfrac{4}{y^2+6y+9} + \dfrac{y}{y^2+y-6}$

1. _____

2. _____

3. _____

4. _____

5. _____

6. _____

7. _____

8. _____

9. _____

10. _____

11. _____

12. _____

13. _____

14. _____

15. _____

16. _____

Challenge

17. Add: $\dfrac{2}{5\Delta^2-20} + \dfrac{\Delta}{10\Delta^3-40\Delta}$

18. Subtract: $\dfrac{100\Delta-1}{100-\Delta^2} - \dfrac{1-100\Delta}{\Delta^2-100}$

17. _____

18. _____

Name:
Instructor:
Date:
Section:

6.5 Complex Fractions

> **Objectives**
> 1. Simplify complex fractions by simplifying numerator and denominator.
> 2. Simplify complex fractions using multiplication first to clear fractions.
>
> **Key Vocabulary**
> complex fraction, main fraction bar

1 Simplify complex fractions by simplifying numerator and denominator.

Example 1 Simplify by simplifying the numerator and the denominator.

a) $\dfrac{\frac{1}{4}+\frac{5}{8}}{\frac{3}{4}}$
b) $\dfrac{\frac{3}{5}-\frac{2}{3}}{\frac{1}{3}+\frac{4}{5}}$
c) $\dfrac{1-\frac{1}{2}}{\frac{2}{7}-1}$

Example 2 Simplify by simplifying the numerator and the denominator.

a) $\dfrac{\frac{1}{4x}+\frac{5}{8y}}{\frac{7}{2y}}$
b) $\dfrac{\frac{x+1}{5x-10}-\frac{3}{10}}{\frac{3}{x-2}+\frac{1}{5x}}$
c) $\dfrac{a-\frac{4}{b}}{\frac{8}{a}-b}$

2 Simplify complex fractions using multiplication first to clear fractions.

Example 3 Simplify the complex fractions in Example 1 by using multiplication first to clear fractions.

a) $\dfrac{\frac{1}{4}+\frac{5}{8}}{\frac{3}{4}}$
b) $\dfrac{\frac{3}{5}-\frac{2}{3}}{\frac{1}{3}+\frac{4}{5}}$
c) $\dfrac{1-\frac{1}{2}}{\frac{2}{7}-1}$

Example 4 Simplify the complex fractions in Example 2 by using multiplication first to clear fractions.

a) $\dfrac{\frac{1}{4x}+\frac{5}{8y}}{\frac{7}{2y}}$
b) $\dfrac{\frac{x+1}{5x-10}-\frac{3}{10}}{\frac{3}{x-2}+\frac{1}{5x}}$
c) $\dfrac{a-\frac{4}{b}}{\frac{8}{a}-b}$

Answers: 1a) $\dfrac{7}{6}$ b) $-\dfrac{1}{17}$ c) $-\dfrac{7}{10}$ 2a) $\dfrac{5x+2y}{28x}$ b) $\dfrac{8x-x^2}{32x-4}$ c) $\dfrac{4a-a^2 b}{ab^2-8b}$ 3. Same answers as Example 1
4. Same answers as Example 2

Name:
Instructor:

Date:
Section:

Practice Set 6.5

Simplify.

1. $\dfrac{\dfrac{1}{a}+\dfrac{3}{b}}{9-\dfrac{3}{b}}$

2. $\dfrac{\dfrac{2}{3}-4}{\dfrac{3}{4}+1}$

1. _____

2. _____

3. $\dfrac{\dfrac{7}{x^2}-\dfrac{3}{x}}{\dfrac{14}{x}}$

4. $\dfrac{\dfrac{3}{x+1}+\dfrac{4}{x-1}}{7x+1}$

3. _____

4. _____

5. $\dfrac{\dfrac{8}{y-2}+\dfrac{y}{2y-4}}{\dfrac{1}{2}+\dfrac{4}{y-2}}$

6. $\dfrac{\dfrac{3}{2x+10}+\dfrac{6}{x+5}}{\dfrac{7}{x+5}}$

5. _____

6. _____

7. $\dfrac{\dfrac{1}{a}-\dfrac{1}{b}}{\dfrac{1}{b}-\dfrac{1}{a}}$

8. $\dfrac{\dfrac{8}{c-3}+\dfrac{2}{3-c}}{\dfrac{12}{c-3}}$

7. _____

8. _____

9. $\dfrac{y}{\dfrac{2y}{x}-\dfrac{3x}{y}}$

10. $\dfrac{\dfrac{x}{y}-\dfrac{y}{x}}{\dfrac{x+y}{y}}$

9. _____

10. _____

11. $\dfrac{\dfrac{m^2}{n}-n}{\dfrac{n^2}{m}-m}$

12. $\dfrac{\dfrac{x-5}{x+5}+\dfrac{x+5}{x-5}}{\dfrac{5}{x^2-5}}$

11. _____

12. _____

Challenge

13. $\dfrac{a^{-1}+b^{-1}}{a^{-1}-b^{-1}}$

14. $\dfrac{\dfrac{1}{x}-\dfrac{1}{y}}{\dfrac{1}{y}-\dfrac{1}{x}}$

13. _____

14. _____

Name:
Instructor:

Date:
Section:

6.6 Solving Rational Equations

Objectives
1. Solve rational equations with integer denominators.
2. Solve rational equations where a variable appears in a denominator.

Key Vocabulary
rational equation, extraneous roots (extraneous solutions)

1 Solve rational equations with integer denominators.

Example 1 Find the LCD of the following fractions.

a) $\dfrac{x}{6}, \dfrac{2}{3}, \dfrac{x}{5}$

b) $\dfrac{2y}{9}, \dfrac{1}{4}, \dfrac{y}{6}, \dfrac{2}{3}$

c) $\dfrac{x-5}{3}, \dfrac{x+1}{4}$

Example 2 Solve each equation and check your answer.

a) $\dfrac{x}{6} - \dfrac{2}{3} = \dfrac{x}{5}$

b) $\dfrac{2y}{9} - \dfrac{1}{4} = \dfrac{y}{6} + \dfrac{2}{3}$

c) $\dfrac{x-5}{3} = \dfrac{x+1}{4}$

2 Solve rational equations where a variable appears in a denominator.

Example 3 Find the LCD of the following rational expressions.

a) $\dfrac{10}{x}, \dfrac{x}{2}, \dfrac{2}{x}$

b) $\dfrac{6}{r}, \dfrac{2}{3r}, \dfrac{8}{6r}$

c) $\dfrac{y}{3-y}, \dfrac{-2y}{2y-1}$

d) $\dfrac{1}{x^2}, \dfrac{2+x}{5x^2}, \dfrac{1}{2x}$

Example 4 Solve each equation and check your answer.

a) $\dfrac{10}{x} - \dfrac{x}{2} = \dfrac{2}{x}$

b) $\dfrac{6}{r} + \dfrac{2}{3r} = \dfrac{8}{6r}$

c) $\dfrac{y}{3-y} = \dfrac{-2y}{2y-1}$

d) $\dfrac{1}{x^2} + \dfrac{2+x}{5x^2} = \dfrac{1}{2x}$

Answers: 1a) 30 b) 36 c) 12   2a) -20 b) $\dfrac{33}{2}$ c) 23   3a) $2x$ b) $6r$ (or $3r$ if $\dfrac{8}{6r}$ is reduced first) c) $(3-y)(2y-1)$ d) $10x^2$ 4a) 4, -4 b) no solution c) 0 d) $\dfrac{14}{3}$

Name:
Instructor:

Date:
Section:

Practice Set 6.6

Solve each equation and check your solution.

1. $\dfrac{x}{2} - \dfrac{21}{10} = \dfrac{x}{2}$

2. $\dfrac{y}{36} + \dfrac{2}{9} = \dfrac{y-2}{18}$

3. $\dfrac{3}{2} - \dfrac{r+4}{6} = \dfrac{r+1}{6}$

4. $\dfrac{10}{x+2} = \dfrac{2}{x-4}$

5. $\dfrac{y-1}{y-4} = \dfrac{3}{y-4}$

6. $\dfrac{4}{3x} - \dfrac{1}{6} = \dfrac{1}{2x}$

7. $\dfrac{x-3}{x+3} = \dfrac{x+3}{x-3}$

8. $\dfrac{x}{5} + \dfrac{4}{x} = -\dfrac{9}{5}$

9. $\dfrac{3}{c-3} - \dfrac{3}{2} = \dfrac{c}{c-3}$

10. $\dfrac{1}{y+3} + \dfrac{4}{y+2} = \dfrac{3y+1}{y^2+5y+6}$

11. $\dfrac{7x}{x^2-9} - \dfrac{5}{x-3} = \dfrac{1}{x+3}$

12. $\dfrac{n}{2n+2} + \dfrac{n}{n+1} = \dfrac{2n-3}{3n+3}$

1. _____

2. _____

3. _____

4. _____

5. _____

6. _____

7. _____

8. _____

9. _____

10. _____

11. _____

12. _____

Challenge

Solve each equation for x.

13. $\dfrac{\dfrac{1}{x}+3}{\dfrac{2}{x}} = 4$

14. $\dfrac{1}{R_1} + \dfrac{1}{R_2} + \dfrac{1}{R_3} = \dfrac{1}{x}$

13. _____

14. _____

Name:
Instructor:

Date:
Section:

6.7 Rational Equations: Applications and Problem Solving

Objectives
1. Set up and solve applications containing rational expressions.
2. Set up and solve motion problems.
3. Set up and solve work problems.

1 Set up and solve applications containing rational expressions.

For each application, set up a rational equation to solve the problem.

Example 1 Dimensions of a Box A box has a length that is one ft less than twice its width. It has a height of $\frac{1}{4}$ ft. If the volume of the box is $\frac{3}{4}$ ft^3, find the length and width of the box.

Example 2 Reciprocals One number is $\frac{5}{6}$ less than its reciprocal. Find all such numbers.

2 Set up and solve motion problems.

Example 3 Great River Road Every Sunday, Matt and his friends travel along the Great River Road in Illinois. One day, Matt wanted to bike while his friends wanted to walk. Matt's speed is 15 mph while his friends' speed is 4 mph, and both travel the same distance. If Matt finishes his ride 1.1 hours before his friends finish their walk, how far did they travel along the road?

Example 4 Kayaking A landscaper by day, Geoffrey kayaks every weekend that he can on the Current River. The speed of the river was 3 mph. If it took Geoffrey the same amount of time to travel 5 mi up river as 14 mi down river, determine the speed at which Geoffrey's kayak would travel in still water.

3 Set up and solve work problems.

Example 5 Homemade Brew A local brew club meets every month to make 10 gal of homemade beer. At one point in the process, one hose fills a barrel while another hose empties the barrel. By itself, the first hose takes 10 min to fill the barrel, and the second hose takes 15 min to empty it. How long will it take to fill an empty barrel?

Answers: 1. $1\frac{1}{2}$ ft and 2 ft 2. $-\frac{3}{2}, \frac{2}{3}$ 3. 6 mi 4. $6\frac{1}{3}$ mph 5. 30 min

Name:
Instructor:

Date:
Section:

Practice Set 6.7

1. Complete the following table given that $d = rt$.

	Rate	Time	Distance
Paul walking	$x - 2$		5
Paul running	x		7

2. Complete the table given only the following information: Mary jogs 5 miles, then walks an additional 2 miles. Her jogging speed is twice her running speed. If the total time Mary spends jogging and walking is 1.125 hours, what is her walking speed?

	Rate	Time	Distance
Mary jogging			
Mary walking			

3. Refer to Problem 2 above. Answer the question.

   3. _____

4. Refer to Problems 2 and 3 above. For how long did Mary walk?

   4. _____

5. **Antique Painting** There is an old painting at an antique mall with an area of 20 ft². The width of the painting is 3 feet less than half its length. Determine the length and width of the painting.

   5. _____

   _____

6. **Reciprocals** One number is 4 times another. The sum of their reciprocals is $\frac{5}{4}$. Determine the numbers.

   6. _____

7. **Canoeing** Anthony and Danielle George are canoeing on the Buffalo River. The speed of the current that day was 4 mph. If it took them the same amount of time to travel 8 mi downstream as 3 mi upstream, determine the speed at which their canoe would travel in still water.

   7. _____

Practice Set 6.7

8. **Stained Glass Project** Dennis is creating a triangular piece of stained glass with a height that is 2 ft more than twice its base. The area is 30 ft². What is the height of the piece of glass?

8. _____

9. **Editing a Manuscript** Carrie and Kristi are editing a manuscript that is just a few days from being printed. If it takes Carrie 4 hours to edit one chapter and it takes Kristi 5 hours to edit the same chapter, how long would it take them to edit the chapter if they worked together?

9. _____

10. **Canoeing** Brian and Tracy are canoeing the Snake River. The speed of the current is 3 mph. If they canoed 18 mi downstream in the time it would take them to canoe 5 mi upstream, how fast were they going downstream?

10. _____

11. **Summer Kitchen** Tom is helping Hank renovate a 120-year-old summer kitchen. Working alone, it takes Hank 8 hours to hang a metal roof on the building. If Hank and Tom work together, it would take them 5 hours to hang the roof. How long would it take Tom to hang the roof if he worked by himself?

11. _____

Name:
Instructor:

Date:
Section:

6.8 Variation

Objectives
1. Set up and solve direct variation problems.
2. Set up and solve inverse variation problems.

Key Vocabulary
variation equations, direct variation, inverse variation, constant of proportionality (or variation constant)

1 Set up and solve direct variation problems.

Example 1 Direct Variation Problem x varies directly as y. If $x = 72$ when $y = 10$, find x when $y = 6$.

Example 2 Gathering Firewood The amount of firewood available, f, at a campsite is directly proportional to the area of the campsite, a. If Alex can find 3 lb of firewood in an 800-square-foot campsite, how large would a campsite need to be in order to find 5 lb of firewood?

2 Set up and solve inverse variation problems.

Example 3 Inverse Variation Problem M varies indirectly as the square of L. If $M = 120$ when $L = 3$, find M when $L = 0.5$.

Example 4 Variation in a Box The area of the bottom of a box, A, varies indirectly as the height of the box, h. If the area of the bottom of a box is 24 square inches when the height is 3 inches, what is the area of the bottom of a box whose height is 6 inches?

Answers: 1. 43.2 2. $1333\frac{1}{3}$ ft^2 3. 4320 4. 12 in.^{2}

Name:
Instructor:

Date:
Section:

Practice Set 6.8

For each scenario, choose **A** if it describes a direct variation or **B** if it describes an inverse variation.

1. The speed of a car and the distance it travels   1. ___

2. The number of people renting a cabin and the cost per person of the rental   2. ___

3. The speed of a car and the amount of time to its destination   3. ___

4. The diameter of a pipe and the amount of water coming out of the pipe   4. ___

5. The number of questions answered correctly on a quiz and the quiz grade   5. ___

Find the quantity indicated.

6. x varies directly as y. Find x when $y = 12$ and $k = 3$.   6. _____

7. M varies indirectly as N. Find M when $N = 40$ and $k = 0.2$.   7. _____

8. A varies directly as the square of B. Find A when $B = 7$ and $k = 4$.   8. _____

9. T varies directly as S. If $T = 5$ when $S = 20$, find T when $S = 40$.   9. _____

10. X varies indirectly as Y. If $X = 0.6$ when $Y = 2$, find X when $Y = 5$. 10. _____

11. E varies indirectly as the square of F. If $E = 8$ when $F = 3$, find E when $F = 9$.   11. _____

Practice Set 6.8

Problem Solving

12. The time it takes to reupholster a chair, *t*, is directly proportional to the square footage of the chair, *s*. If it takes 9 hours to reupholster a chair that is 6.5 ft², how long will it take to reupholster a chair that is 11 ft²?

12. _____

13. The number of toothbrush rugs, *r*, that are made at Shannondale Native Craft Workshop is directly proportional to the amount of tuition, *t*, paid by registrants. If registrants paid a total of $2045 and 38 toothbrush rugs were made, how many rugs were made when registrants only paid $1230?

13. _____

14. The time, *t*, it takes a newspaper delivery truck to deliver a specific number of newspapers is inversely proportional to the number of newspaper delivery people, *n*, working a route. If it takes 1.5 hours for 4 people to deliver some papers on a 12-mile route, how long will it take 3 people to deliver the same number of papers on the same route?

14. _____

15. The amount of liquid, *l*, in a can is directly proportional to its diameter, *d*. If there is 64 fl oz of broth in a can with a diameter of 4.5 in., what is the diameter of a can that can hold 20 fl oz of broth?

15. _____

16. The volume of a gas, V, varies inversely as its pressure, P. If the volume is 500 cubic centimeters while the pressure is 300 millimeters (mm) of mercury, find the volume when the pressure is 40 mm of mercury.

16. _____

Chapter 6 Vocabulary Reference Sheet

Term	Definition	Example
rational expression	An expression of the form $\frac{p}{q}$, where p and q are polynomials and $q \neq 0$	$\frac{x+2}{x+5}$
simplify	To simplify a rational expression, factor both numerator and denominator as completely as possible and then divide out any factors common to the numerator and denominator.	$\frac{2x-4}{x-2} = \frac{2(x-2)}{x-2} = 2$
LCD	The LCD (least common denominator) of two or more rational expressions is the LCM (least common multiple) of their denominators.	The LCD of $\frac{x-4}{x-2}$ and $\frac{4}{x+3}$ is $(x-2)(x+3)$.
complex fraction	A fraction that has a fraction in its numerator or its denominator or in both its numerator and denominator	$\frac{\frac{x}{3} - \frac{6}{x}}{\frac{5}{2x}}$
rational equation	An equation that contains one or more rational expressions	$\frac{4}{x} = \frac{3}{x} + 7$ is a rational equation.
direct variation	A variation in which both variables either increase together or decrease together	The amount of money made, m, is directly proportional to the number of hours, h, someone works. This is represented by the equation $m = kh$.
inverse variation	Variation in which one variable increases as the other decreases, and vice versa	The amount of time it takes, t, to reach a destination is indirectly proportional to the speed, r, of the vehicle going to the destination. This is represented by the equation $t = \frac{k}{r}$.
constant of proportionality (or variation constant)	The variation constant that exists in direct and indirect variation problems	The value k in each of the variation examples above is the constant of proportionality. If this value is not given in a variation problem, then it is usually the first value that needs to found.

Notes:

Name:
Instructor:

Date:
Section:

Chapter 6 Practice Test A

Simplify.

1. $\dfrac{2x-10}{x^2-25}$

2. $\dfrac{6y-6}{3y-3y^2}$

1._____

2._____

Determine the value or values of the variable for which the expression is defined.

3. $\dfrac{x-7}{x+1}$

4. $\dfrac{4x-8}{x^2-36}$

3._____

4._____

Perform the indicated operation.

5. $\dfrac{75a^2b}{4a^2bc} \cdot \dfrac{12ab}{25ab^3}$

6. $\dfrac{x^2-25}{4x+20} \div \dfrac{3x-15}{12x+12}$

5._____

6._____

7. $\dfrac{x^2+2x-3}{x^2-1} \cdot \dfrac{4x^2-1}{2x^2+5x-3}$

8. $\dfrac{x^2+3x+9}{x^3-27} \div \dfrac{4x+12}{9-x^2}$

7._____

8._____

9. $\dfrac{6x+1}{12x} + \dfrac{4x+5}{12x}$

10. $\dfrac{2x}{x-4} + \dfrac{8}{4-x}$

9._____

10._____

11. $\dfrac{5}{3x^2y} - \dfrac{11y}{6x}$

12. $\dfrac{1}{7x-7} - \dfrac{x+2}{14x^2-14}$

11._____

12._____

Practice Test 6A

13. $\dfrac{6}{x^2+x-6}+\dfrac{x}{x^2-4}$

14. $\dfrac{2x}{x+1}-\dfrac{4}{x+3}$

15. $\dfrac{2+\dfrac{2}{5}}{1-\dfrac{7}{10}}$

16. $\dfrac{\dfrac{4}{x}-\dfrac{1}{y}}{\dfrac{11}{y}}$

Solve.

17. $\dfrac{5x}{12}+\dfrac{4}{3}=\dfrac{1}{2}$

18. $4+\dfrac{5}{x}=9$

19. $\dfrac{1}{16}+\dfrac{1}{2x}=\dfrac{-1}{x^2}$

20. $\dfrac{r}{r-3}-3=\dfrac{3}{r-3}$

13. _____

14. _____

15. _____

16. _____

17. _____

18. _____

19. _____

20. _____

Solve.

21. **Working Together** When working together, it takes Brianne and her sister, Kayla, 45 min to clean their swimming pool. How long will it take Kayla to clean the pool alone if it takes Brianne 2 hours to clean the pool by herself?

22. **Determine a Number** The sum of a number and its reciprocal is equal to $\dfrac{17}{4}$. Find the number.

23. **Landscaping** The time it takes, t, for a landscaper to weed a garden is directly proportional to the garden's area, a. If a 24 ft² garden takes 2.5 hours to weed, how long will it take to weed a garden that has an area of 40 ft²?

21. _____

22. _____

23. _____

Name:
Instructor:

Date:
Section:

Chapter 6 Practice Test B

1. Determine the value or values that make $\dfrac{2x}{9x-3}$ undefined.

   a) -3 b) 0 c) $\dfrac{1}{3}$ d) 3

2. Determine the value or values that make $\dfrac{x-4}{x^2-1}$ undefined.

   a) -4 b) 1 c) $-1, 1$ d) 4

3. Simplify $\dfrac{7x-21}{6x^2-18x}$.

   a) 0 b) $\dfrac{7}{x}$ c) $\dfrac{7x-21}{6x}$ d) $\dfrac{7}{6x}$

4. Simplify $\dfrac{8y^2-12y}{6-4y}$.

   a) $2y$ b) $-2y$ c) $\dfrac{4y^2-6y}{3-2y}$ d) $\dfrac{4y^2-3}{2}$

5. Multiply $\dfrac{5x-10}{25xy^2}\cdot\dfrac{10x^2y}{x^2-4}$.

   a) $\dfrac{105}{2y}$ b) $\dfrac{10x}{5y(x+2)}$ c) $\dfrac{2x}{y(x+2)}$ d) $\dfrac{2}{x+2}$

6. Divide $\dfrac{15-3x}{8x+8}\div\dfrac{2x^3-50x}{4x^3+4x^2}$.

   a) $\dfrac{-4(x+5)}{3x}$ b) $\dfrac{-3x}{4(x+5)}$ c) $\dfrac{3x}{4(x+5)}$ d) $\dfrac{5}{32x}$

7. Subtract $\dfrac{3x+7}{x-2}-\dfrac{2x+9}{x-2}$.

   a) $\dfrac{x+16}{x-2}$ b) 8 c) -8 d) 1

8. Add $\dfrac{2}{3x}+\dfrac{9}{4y}$.

   a) 2 b) $\dfrac{11}{7xy}$ c) $\dfrac{2}{xy}$ d) $\dfrac{27x+8y}{12xy}$

Practice Test 6B

9. Add $\dfrac{12}{4-y} + \dfrac{3y}{y-4}$.

   a) 3
   b) −3
   c) $\dfrac{-3(y+4)}{(y+4)(y-4)}$
   d) $\dfrac{3(y+4)}{(y+4)(y-4)}$

10. Add $\dfrac{1}{x^2-x-6} + \dfrac{x}{2x-6}$.

    a) $\dfrac{x^3+x^2-4x-6}{2(x-2)(x+3)(x-3)}$
    b) $\dfrac{-x^3-x^2+4x+6}{2(x-2)(x+3)(x-3)}$
    c) $\dfrac{-x^2-2x-2}{2(x+2)(x-3)}$
    d) $\dfrac{x^2+2x+2}{2(x+2)(x-3)}$

11. Simplify $\dfrac{3-\dfrac{2}{3}}{1+\dfrac{5}{6}}$.

    a) $\dfrac{14}{11}$
    b) $-\dfrac{14}{11}$
    c) $\dfrac{6}{11}$
    d) $-\dfrac{6}{11}$

12. Simplify $\dfrac{\dfrac{4}{a}-\dfrac{6}{b}}{\dfrac{5}{ab}+a}$.

    a) $\dfrac{4b-6a}{5+a^2b}$
    b) $\dfrac{6a-4b}{5+a^2b}$
    c) $\dfrac{2ab}{5+ab}$
    d) $\dfrac{-2ab}{5+ab}$

13. Solve $\dfrac{x}{12} + \dfrac{2}{3} = \dfrac{1}{4}$.

    a) $-\dfrac{1}{5}$
    b) −5
    c) −1
    d) no solution

14. Solve $\dfrac{y}{y-5} - 5 = \dfrac{5}{y-5}$.

    a) 5
    b) −5
    c) 10
    d) no solution

15. **Area of a Rectangle** The height of a triangular-shaped antique sign is 2 ft less than the length of its base. If the area of the sign is $7\dfrac{1}{2}$ ft², find the sign's height.

    a) 2 ft
    b) 3 ft
    c) 5 ft
    d) 5.5 ft

16. The time it takes for a student to graduate, t, is inversely proportional to the number of credit hours, c, that she completes each semester. A student completing 15 credit hours per semester takes 8 semesters to graduate. How long will it take a student to graduate if she takes 10 credit hours per semester?

    a) 11 semesters
    b) 12 semesters
    c) 13 semesters
    d) 14 semesters

Name:
Instructor:

Date:
Section:

7.1 The Cartesian Coordinate System and Linear Equations in Two Variables

Objectives
1. Plot points in the Cartesian coordinate system.
2. Determine whether an ordered pair is a solution to a linear equation.

Key Vocabulary
graph, Cartesian (rectangular) coordinate system, quadrants, y-axis, x-axis, origin, ordered pair, x-coordinate, y-coordinate, linear equation in two variables, standard form, graph, collinear

1 Plot points in the Cartesian coordinate system.

Example 1 Plot the following points on the same set of axes. Label each point with its corresponding letter.

a) $(2, 3)$ b) $(4, -3)$ c) $(-4, -3)$

d) $(-4, 2)$ e) $(-3, 0)$ f) $(0, 3)$

Example 2 List the ordered pairs for each point shown and indicate the quadrant in which each point lies or the axis on which it lies..

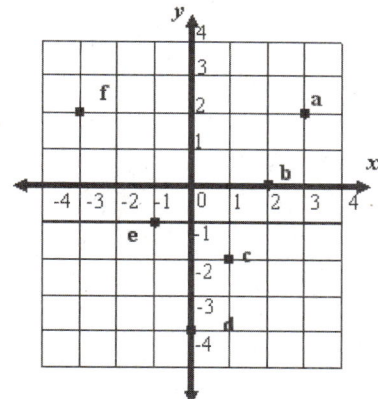

Answers: 1.

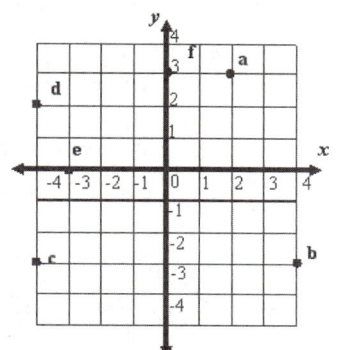

2a) $(3,2)$; I   b) $(2,0)$; x-axis   c) $(1, -2)$; IV
d) $(0, -4)$; y-axis   e) $(-1, -1)$; III   f) $(-3, 2)$; II

Examples 7.1

2 Determine whether an ordered pair is a solution to a linear equation.

Example 3 Determine which ordered pairs satisfy the given equation.

a) $y = 2x - 1$;
$(3, 5), (-2, 3), (0, -1)$

b) $3x - 4y = 5$;
$(2, -1), (7, 4), (-1, -2)$

c) $x + 2y = -3$;
$\left(0, -\dfrac{3}{2}\right), (5, -4), \left(\dfrac{1}{3}, -\dfrac{2}{3}\right)$

d) $\dfrac{4}{3}x - 2y = 8$;
$(-6, 5), (-3, -6), \left(1, -\dfrac{10}{3}\right)$

Example 4 For each of the following ordered pairs, determine the unknown coordinate that makes the point a solution to the equation $x - 3y = 5$.

a) $(x, 2)$
b) $(-4, y)$
c) $(x, 0)$
d) $(-3, y)$

Answers: 3a) $(3, 5), (0, -1)$ b) $(7, 4), (-1, -2)$ c) $\left(0, -\dfrac{3}{2}\right), (5, -4)$ d) $(-3, -6), \left(1, -\dfrac{10}{3}\right)$ 4a) 11 b) -3 c) 5 d) $-\dfrac{8}{3}$

Name:
Instructor:

Date:
Section:

Practice Set 7.1

Determine the quadrant in which the point lies or the axis on which it lies.

1. $(-3, 2)$
2. $(-8, -10)$

1. _____

2. _____

3. $(-32, 56)$
4. $(5, 6)$

3. _____

4. _____

5. $\left(\dfrac{1}{2}, -5\right)$
6. $(8, 0)$

5. _____

6. _____

7. Determine the ordered pair for each point on the set of axes below.

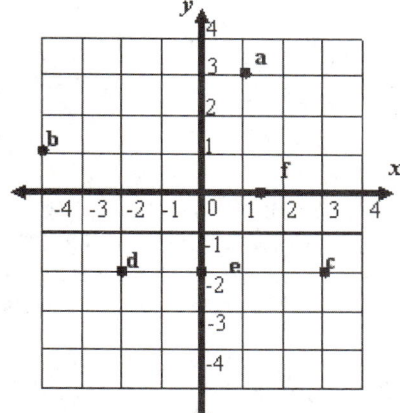

7. a) _____

b) _____

c) _____

d) _____

e) _____

f) _____

Plot each point on the set of axes below. Label the point with its corresponding letter.

8. a) $(0, 3)$ b) $(-4, -2)$ c) $(-3, 0)$ d) $\left(\dfrac{3}{2}, 4\right)$

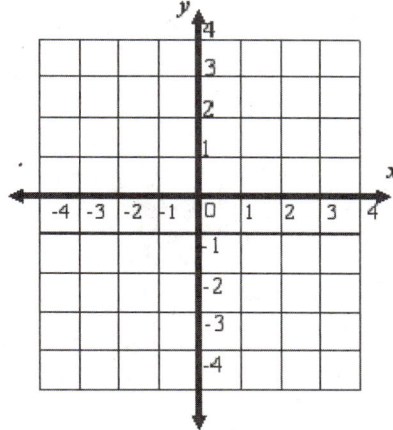

Practice Set 7.1

Plot each point on the set of axes below. Label the point with its corresponding letter.

9. a) $\left(\dfrac{1}{2}, 3\right)$ b) $(3, -3)$ c) $(0, -4)$ d) $\left(-4, -\dfrac{7}{3}\right)$

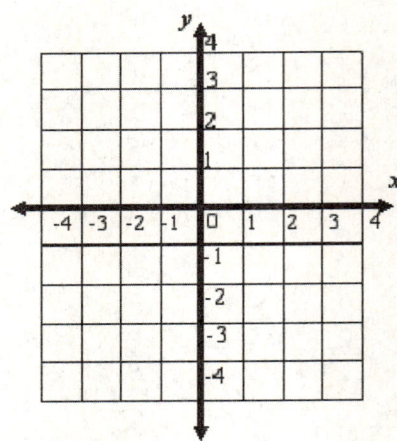

Determine which of the given points satisfy the given equation.

10. $y = 2x - 5$; $(3, 1), (-2, -9), (5, 7)$ 10. _____

11. $3x + 2y = 12$; $(1, 3), (2, 3), (4, 0)$ 11. _____

Determine the unknown coordinate for each given point so that each point satisfies the given equation.

12. $y = 6x - 2$; $(3, y), (x, 10)$ 12. _____

_____

13. $y = \dfrac{3}{2}x + 4$; $(-2, y), (x, 7)$ 13. _____

_____

14. $3x + 4y = -6$; $(0, y), (x, -3)$ 14. _____

_____

Challenge

15. Consider all points in the second quadrant. 15. a) _____
    a) For these points, what do all of the x-coordinates have in common?
    b) What do all of the y-coordinates have in common?

b) _____

Name:
Instructor:

Date:
Section:

7.2 Graphing Linear Equations

Objectives

1. Graph linear equations by plotting points.
2. Graph linear equations of the form $ax + by = 0$.
3. Graph linear equations using the x- and y-intercepts.
4. Graph horizontal and vertical lines.
5. Study applications of graphs.

Key Vocabulary

x-intercept, y-intercept, horizontal line, vertical line

1 Graph linear equations by plotting points.

Example 1 Graph each linear equation by plotting points. Label the points.

a) $y = 2x - 1$

b) $x + 2y = -3$

Answers: 1a)

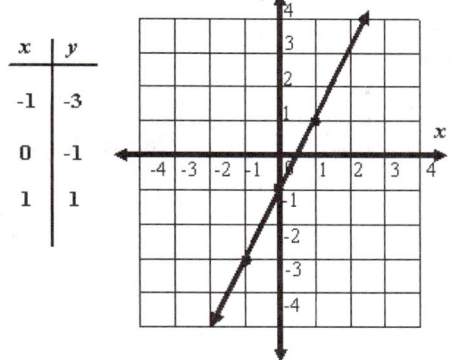

b)

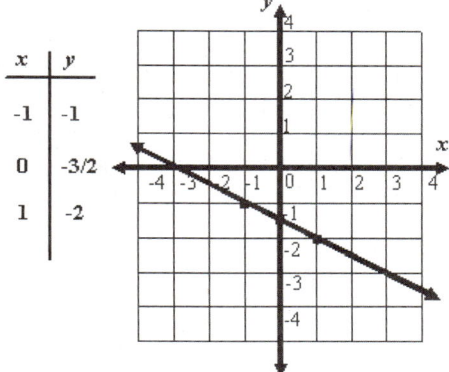

Examples 7.2

2 Graph linear equations of the form $ax + by = 0$.

Example 2 Graph each linear equation of the form $ax + by = 0$. Label the axes accordingly.

a) $2x - y = 0$

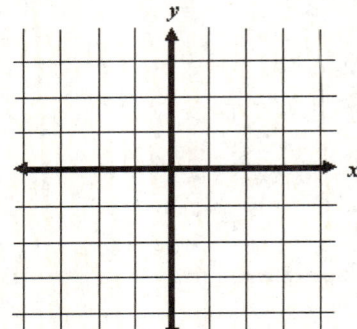

b) $4x + 3y = 0$

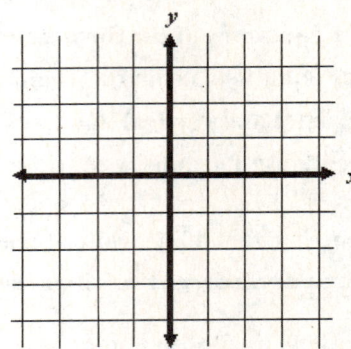

3 Graph linear equations using the *x*- and *y*-intercepts.

Example 3 Graph each linear equation using the *x*- and *y*-intercepts. Label the intercepts.

a) $5x - 3y = 15$

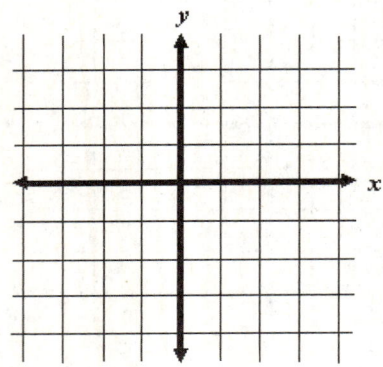

b) $y = 6x - 3$

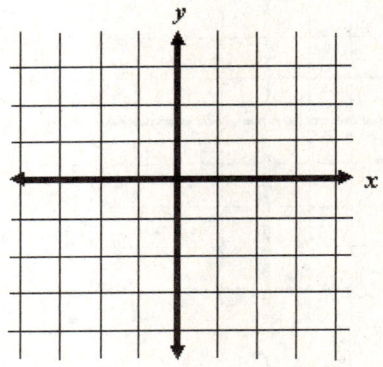

Answers:

2a)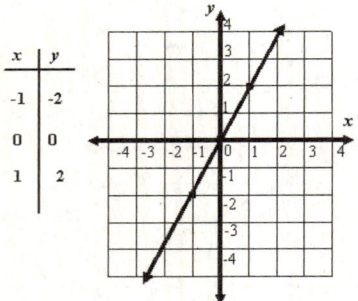

b)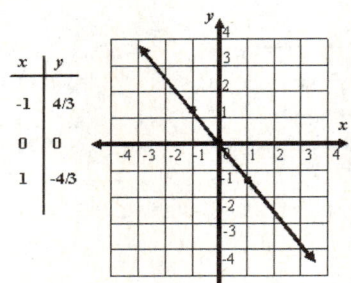

3a)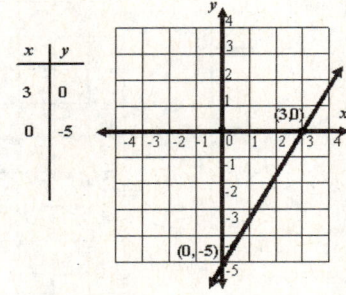

Examples 7.2

4 Graph horizontal and vertical lines.

Example 4 Graph each linear equation. Label the axes accordingly.

a) $y = 3$

b) $x = -2$

5 Study applications of graphs.

Example 5 Family and Friends Video charges $2.50 to rent a new release movie. For each day that the movie is late, the customer is charged $0.50.

a) Write an equation for the total cost, C, of renting a movie if the movie is d days late.

b) Graph the equation for up to and including 5 days late.

c) Use the graph to estimate the total cost for renting a new release movie if the movie is 2 days late.

d) Use the graph to estimate the number of days a new release is late if the total cost is $4.50.

Answers:

3b)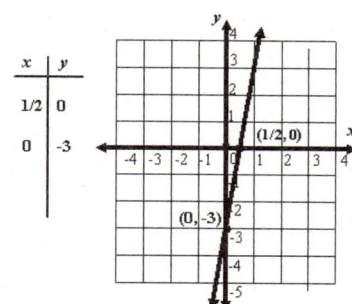

4a)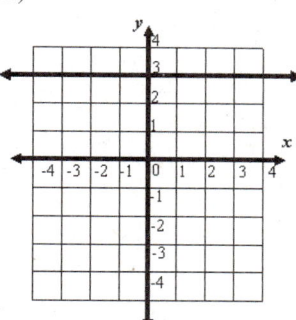

b)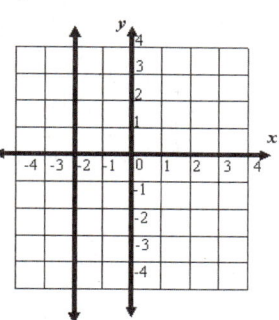

5a) $C = 2.50 + 0.50d$ b) 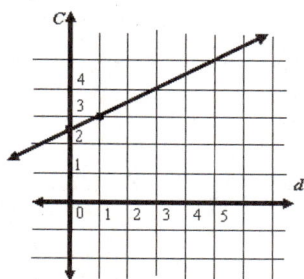   c) $3.50   d) 4 days late

Name:
Instructor:

Date:
Section:

Practice Set 7.2

Graph each linear equation by plotting points. Label the axes accordingly.

1. $y = 5x - 2$

2. $y = -3x + 4$

3. $y = \dfrac{2}{3}x - 2$

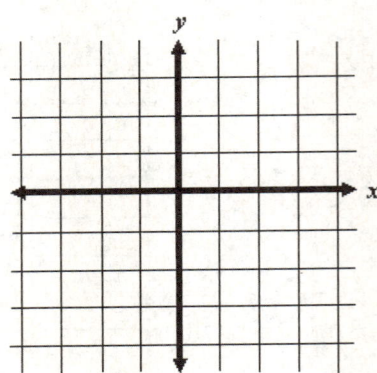
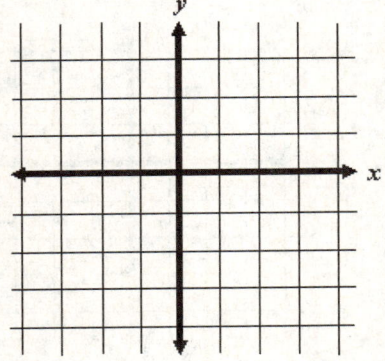

4. $y = -\dfrac{3}{4}x - 5$

5. $3x - 6y = 12$

6. $5x - 2y = -4$

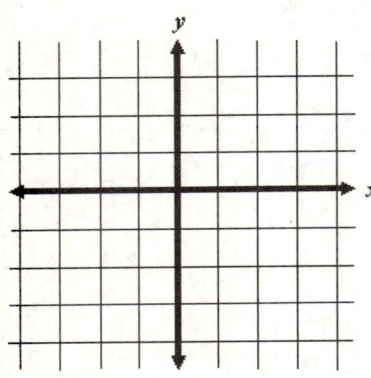
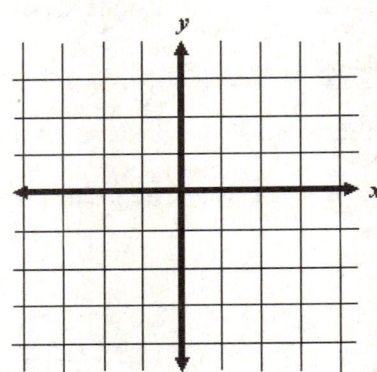
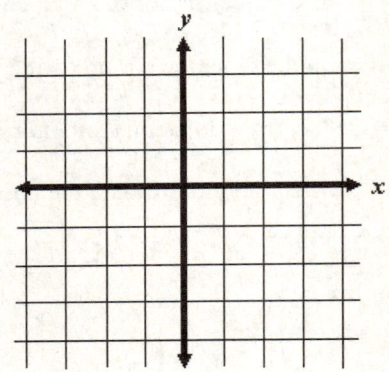

Graph each linear equation by using x- and y-intercepts. Label the intercepts.

7. $2x - 4y = 8$

8. $x - y = 5$

9. $y = 3x - 5$

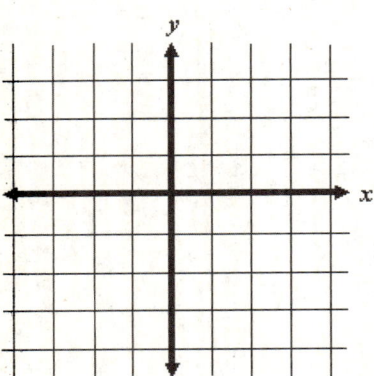
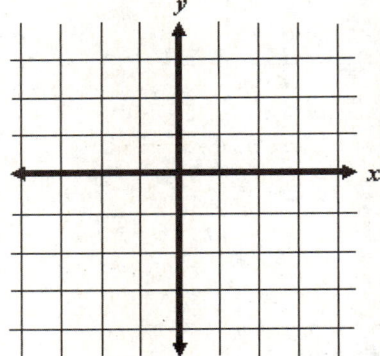
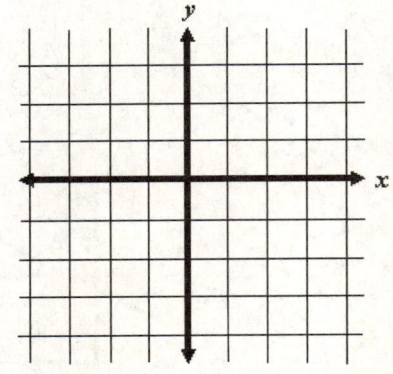

Practice Set 7.2

Graph each linear equation. Label the axes accordingly.

10. $y = -4$

11. $x = \dfrac{5}{2}$

12. $x + 3 = 5$

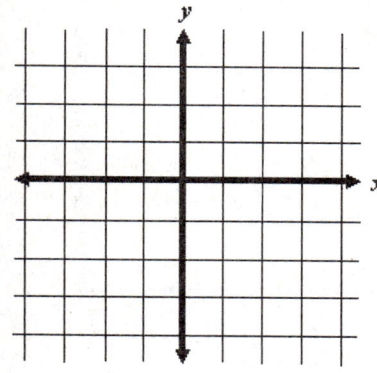 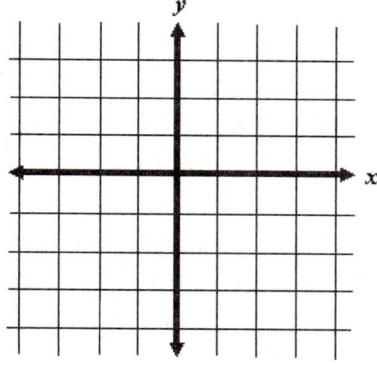 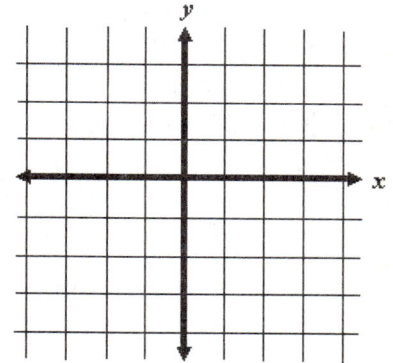

Write the equation represented by the given graph.

13.

14.

13. _____

14. _____

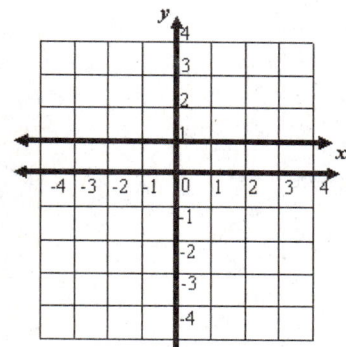 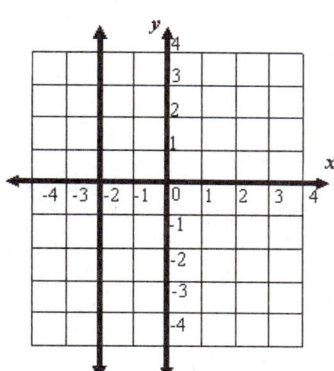

15. The relationship between a temperature in Fahrenheit, F, and Celsius, C, is given by the linear equation $F = \dfrac{9}{5}C + 32$.

   a) Graph the equation with Celsius on the horizontal axis and Fahrenheit on the vertical axis. Include temperatures from $-20°C$ to $20°C$. [Note: Label the axes accordingly.]

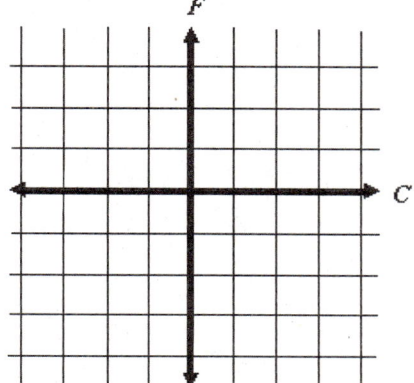

   b) Use the graph to estimate the temperature in Fahrenheit for 15°C.
   c) Use the graph to estimate the temperature in Celsius for 20°F.

15. b) _____

   c) _____

Name:
Instructor:
Date:
Section:

7.3 Slope of a Line

Objectives
1. Find the slope of a line.
2. Recognize positive and negative slopes.
3. Examine the slopes of horizontal and vertical lines.
4. Examine the slopes of parallel and perpendicular lines.

Key Vocabulary
slope, rate of change, parallel lines, perpendicular lines, negative reciprocals

1 Find the slope of a line.

Example 1 Find the slope of each line.

a)

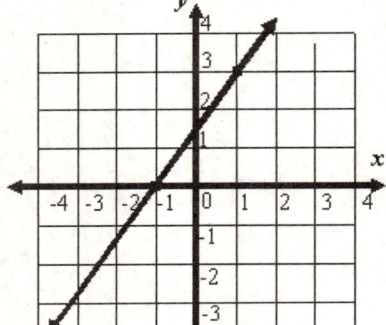

b)

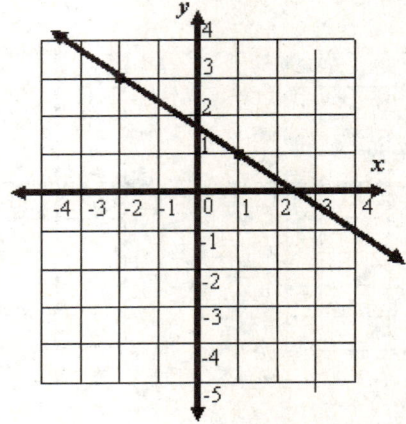

Example 2 Find the slope of the line that passes through the given points.

a) $(1, 2), (4, 5)$ b) $(-3, 4), (2, 1)$ c) $(0, 4), (3, -2)$

Answers: 1a) $\frac{3}{2}$ b) $-\frac{2}{3}$ 2a) 1 b) $-\frac{3}{5}$ c) -2

Examples 7.3

2 Recognize positive and negative slopes.

Example 3 Determine the sign (positive or negative) of the slope of the given line, L1.

a)
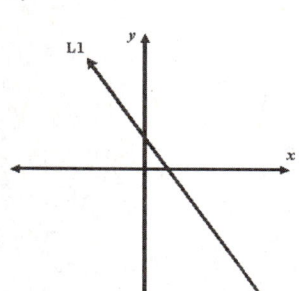

b)
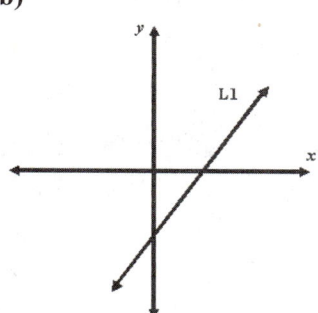

c)
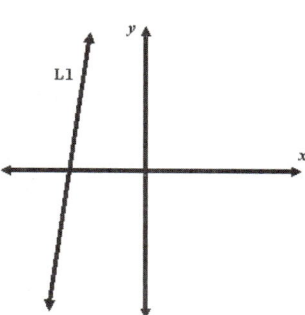

3 Examine the slopes of horizontal and vertical lines.

Example 4 Find the slope given the following information.

a) The line L1

b) The line L2

c) The line L3 through the points $(3, -2)$, $(1, -2)$.

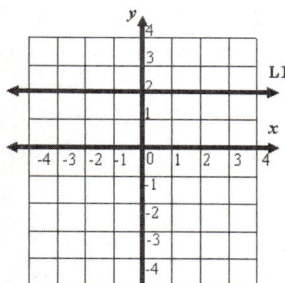
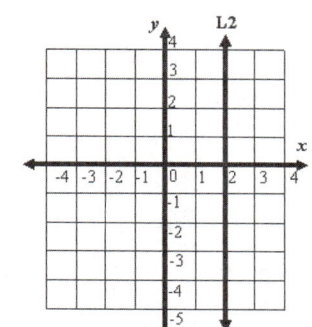
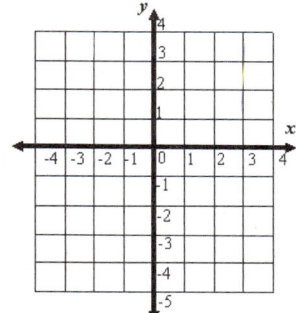

4 Examine the slopes of parallel and perpendicular lines.

Example 5 Are lines L1 and L2 parallel or perpendicular? Find the slopes of L1 and L2 in parts a and b.

a)
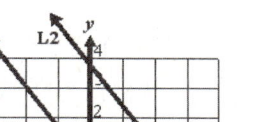
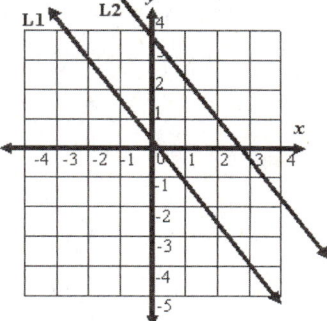

b)
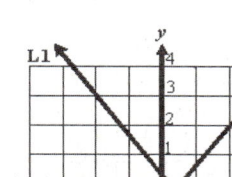
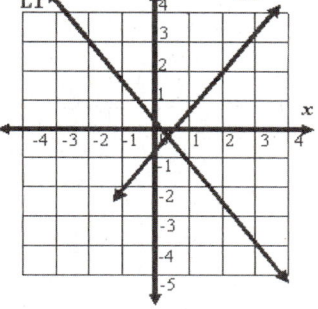

c) L1 has $m = -\dfrac{1}{2}$ and L2 has $m = 2$.

Answers: 3a) negative b) positive c) positive 4a) 0 b) undefined c) 0 5a) parallel; $-\dfrac{7}{5}$ and $-\dfrac{7}{5}$ b) neither; $-\dfrac{7}{5}$ and $\dfrac{4}{3}$ c) perpendicular

Copyright © 2015 by Pearson Education, Inc.   Angel/Runde, *Elementary Algebra for College Students*, 9e

Name:
Instructor:
Date:
Section:

Practice Set 7.3

Determine whether the lines with the description on the left are parallel, perpendicular, or neither.

1. $m = 5$, $m = 5$
2. $m = \dfrac{2}{5}$, $m = -\dfrac{5}{2}$
3. $m = \dfrac{1}{3}$, $m = 3$
4. $m = \dfrac{1}{5}$, $m = -5$

A. parallel
B. perpendicular
C. neither

1. _____
2. _____
3. _____
4. _____

Find the slope of the line through the given pair of points.

5. $(3, 5)$, $(-2, 1)$
6. $(-3, 6)$, $(8, -3)$

5. _____
6. _____

7. $(-4, -5)$, $(4, 5)$
8. $(0, 6)$, $(3, 8)$

7. _____
8. _____

Find the slope of the given line.

9.

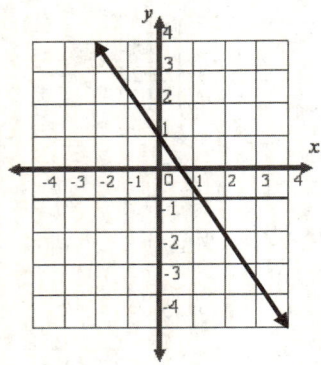

10.

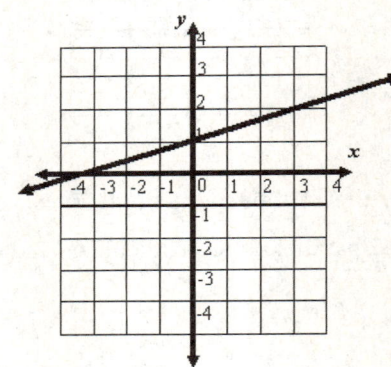

9. _____

10. _____

11.

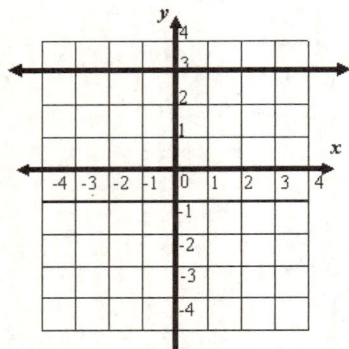

12.

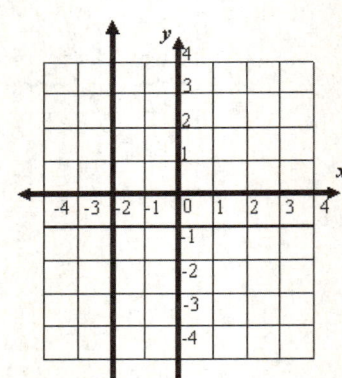

11. _____

12. _____

Practice Set 7.3

Graph the line that goes through the given point and has the given slope. Label the axes accordingly.

13. $(-1, 2)$; $m = \dfrac{1}{3}$

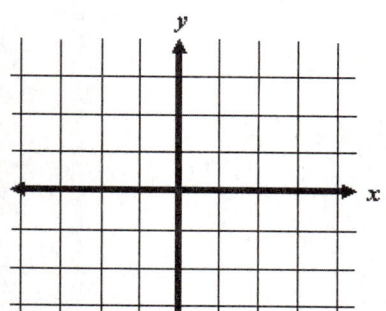

14. $(1, -3)$; $m = -\dfrac{1}{2}$

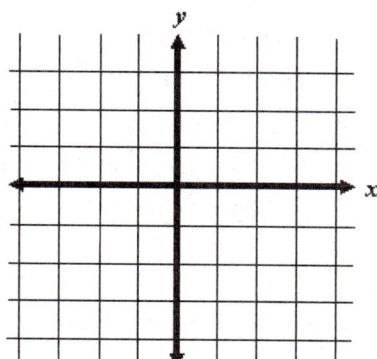

15. $(0, 1)$; $m = 2$

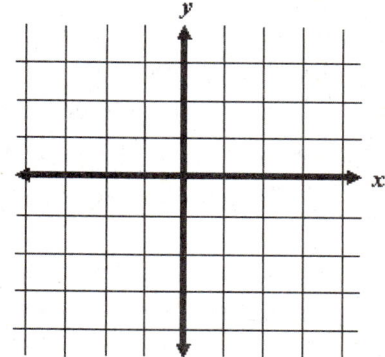

16. $(-3, 1)$; $m = 0$

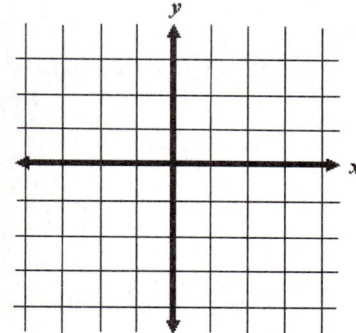

17. $(-3, 1)$; m is undefined

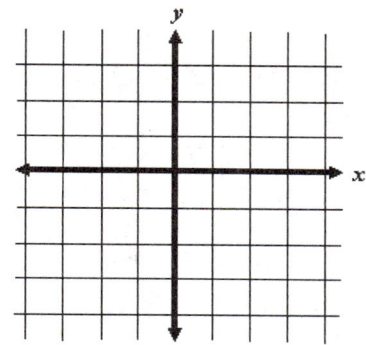

18. The slope of a given line is -3.
    a) What is the slope of a line that is parallel to the given line?
    b) What is the slope of a line that is perpendicular to the given line?

18. a)_____
    b)_____

19. The slope of a given line is $-\dfrac{2}{7}$.
    a) What is the slope of a line that is parallel to the given line?
    b) What is the slope of a line that is perpendicular to the given line?

19. a)_____
    b)_____

20. The slope of a given line is 0.
    a) What is the slope of a line that is parallel to the given line?
    b) What is the slope of a line that is perpendicular to the given line?

20. a)_____
    b)_____

Name:
Instructor:

Date:
Section:

7.4 Slope-Intercept and Point-Slope Forms of a Linear Equation

Objectives
1. Write a linear equation in slope-intercept form.
2. Graph a linear equation using the slope and y-intercept.
3. Use the slope-intercept form to determine the equation of a line.
4. Use the point-slope form to determine the equation of a line.
5. Compare the three methods of graphing linear equations.

Key Vocabulary
slope-intercept form, point-slope form

1 Write a linear equation in slope-intercept form.

Example 1 Write each equation in slope-intercept form, and state the slope and y-intercept.

a) $2x - 3y = 12$

b) $4x - 2y = -6$

Example 2 Determine whether the two equations represent lines that are parallel, perpendicular, or neither.

a) $x - 2y = 12$
$3x - 6y = 18$

b) $4x + y = -1$
$2x - 8y = -16$

2 Graph a linear equation using the slope and y-intercept.

Example 3 Write each equation in slope-intercept form and graph each equation. Label the axes accordingly.

a) $2x - y = 3$

b) $x + 3y = 6$

Answers: 1a) $y = \dfrac{2}{3}x - 4$; $m = \dfrac{2}{3}$; $(0, -4)$ b) $y = 2x + 3$; $m = 2$; $(0, 3)$ 2a) parallel b) perpendicular

3a) $y = 2x - 3$; 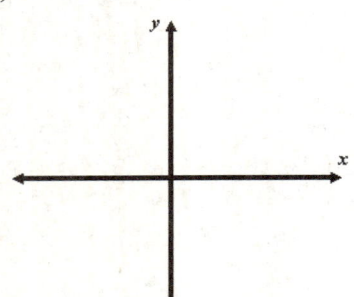   b) $y = -\dfrac{1}{3}x + 2$;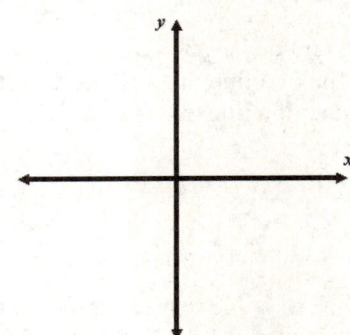

Examples 7.4

3 Use the slope-intercept form to determine the equation of a line.

Example 4 Find the equation in slope-intercept form for the line with y-intercept $(0, -2)$ and slope $\frac{2}{3}$.

Example 5 Find the slope and y-intercept of the line below. Write the equation in slope-intercept form.

a)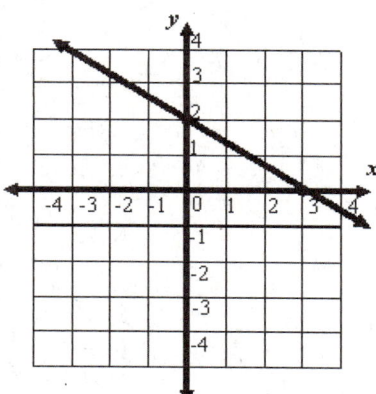

b)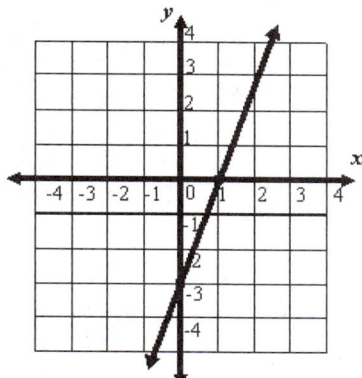

4 Use the point-slope form to determine the equation of a line.

Example 6 Write the equation for the line with slope -3 that goes through the point $(1, -4)$. Write the equation in point-slope form.

Example 7 Write the equation for the line that goes through the points $(-5, 6)$ and $(-2, 3)$. Write the equation in point-slope form.

Answers: 4. $y = \frac{2}{3}x - 2$ 5a) $y = -\frac{2}{3}x + 2$ b) $y = 3x - 3$ 6. $y + 4 = -3(x - 1)$ 7. $y - 6 = -1(x + 5)$ or $y - 3 = -1(x + 2)$

Examples 7.4

Example 8 Write the equation for the line that goes through the point $(-4, 3)$ and is parallel to $2x - 4y = 5$. Write the equation in slope-intercept form.

5 Compare the three methods of graphing linear equations.

Example 9 Graph the linear equation $2x + 3y = 3$ by a) plotting points, b) using x- and y-intercepts, and c) using the slope-intercept method.

a)
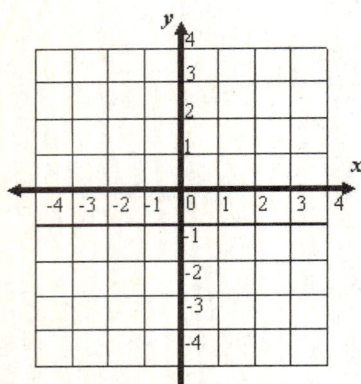

b)
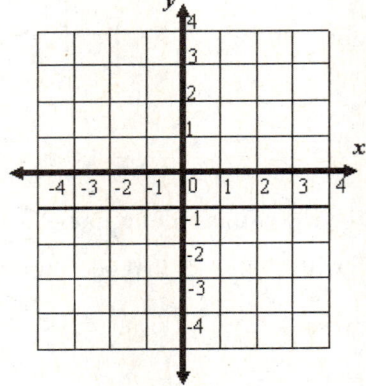

c)
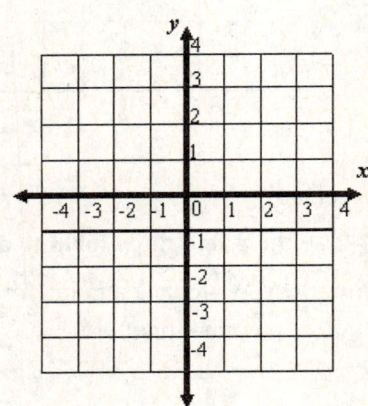

Answers: 8. $y = \dfrac{1}{2}x + 5$ 9a) see table below b) x-intercept $\left(\dfrac{3}{2}, 0\right)$; y-intercept $(0, 1)$ c) $m = -\dfrac{2}{3}; (0, 1)$

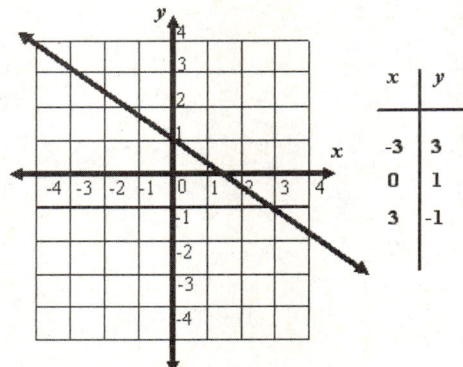

Name:
Instructor:

Date:
Section:

Practice Set 7.4

Find the slope and *y*-intercept of each linear equation.

1. $y = 2x + 3$

2. $y = \dfrac{4}{3}x - 5$

3. $3x - y = 2$

4. $2x + 3y = 6$

1. _____

2. _____

3. _____

4. _____

For each line below, **a)** find the slope and *y*-intercept, and **b)** state the equation of the line in slope-intercept form. (Note: The axes are marked in 1-unit increments.)

5.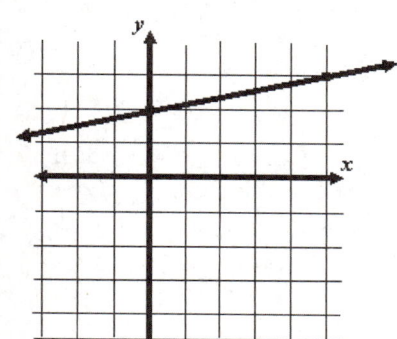

6.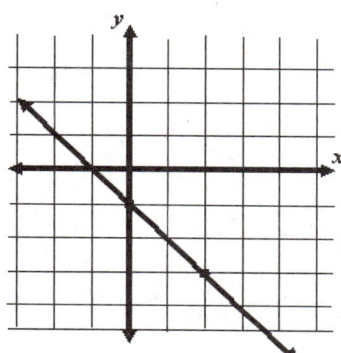

7.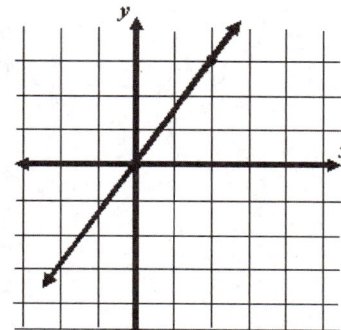

8.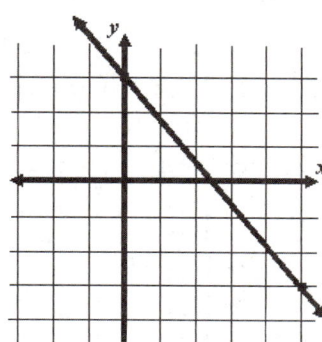

5. a) _____

   b) _____

6. a) _____

   b) _____

7. a) _____

   b) _____

8. a) _____

   b) _____

Determine whether the two equations represent lines that are parallel, perpendicular, or neither.

9. $6x - 12y = 24$
   $3x - 6y = -9$

10. $5y - 4x = 15$
    $5x - 4y = 8$

9. _____

10. _____

Practice Set 7.4

11. $6x + y = 1$
     $2x + 12y = 8$

12. $x = 2$
     $y = -2$

11. _____

12. _____

Write the equation of each line with the given properties. Write the equation in slope-intercept form.

13. through $(0, 5)$, $m = -4$

14. through $(-2, -1)$, $m = -\dfrac{5}{2}$

13. _____

14. _____

15. through $(-8, 12)$, $m = \dfrac{1}{6}$

16. through $(4, 8)$ and $(-2, 6)$

15. _____

16. _____

17. a) Determine the equation of the line that goes through the point $(-2, 1)$ and is parallel to the line $x - y = 4$.
    b) Graph the parallel lines below. Label the axes accordingly.

17. a) _____

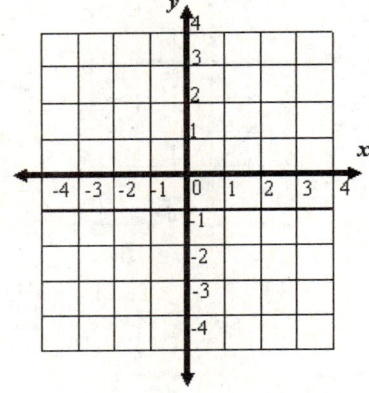

18. a) Determine the equation of the line that goes through the point $(-3, -2)$ and is perpendicular to the line $3x + y = 1$.
    b) Graph the perpendicular lines below. Label the axes accordingly.

18. a) _____

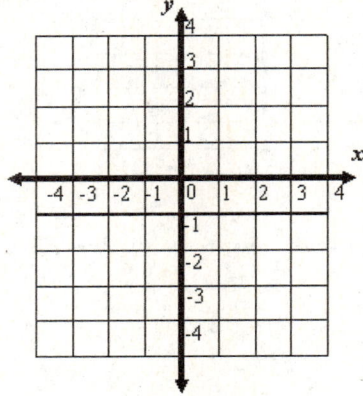

Name:
Instructor:

Date:
Section:

7.5 Graphing Linear Inequalities

Objectives

1 Graph linear inequalities in two variables.

Key Vocabulary
linear inequality, boundary

1 Graph linear inequalities in two variables.

Example 1 Determine if each ordered pair is a solution to the inequality.

a) $y < 3x - 1$; $(0, -2)$; $(-2, 0)$

b) $2x - 3y \leq 9$; $(0, 0)$; $(-4, -8)$

Example 2 Graph each linear inequality. Label the axes accordingly. (Note: a and c are from a and b above.)

a) $y < 3x - 1$

b) $y \geq -4x + 2$

c) $2x - 3y \leq 9$

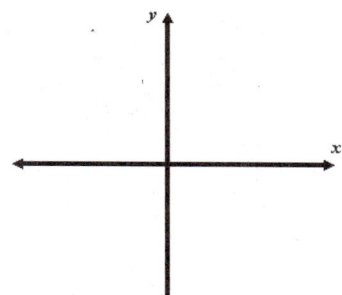 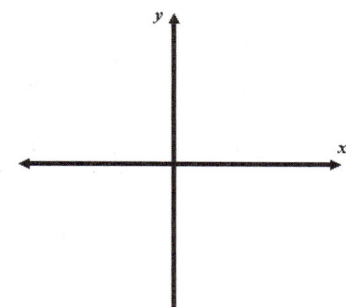 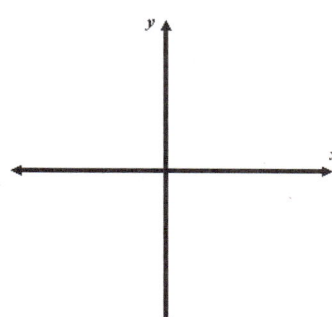

Answers: 1a) $(0, -2)$ is a solution; $(-2, 0)$ is not a solution.   b) $(0, 0)$ is a solution; $(-4, -8)$ is not a solution.

2a)

b)

c)

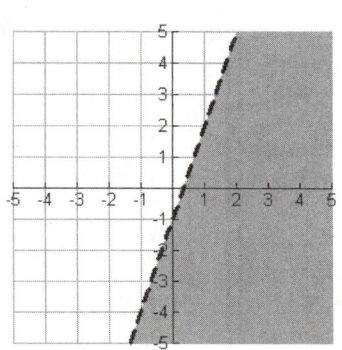 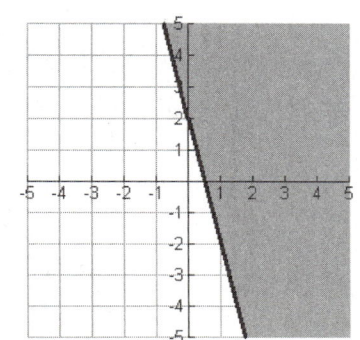 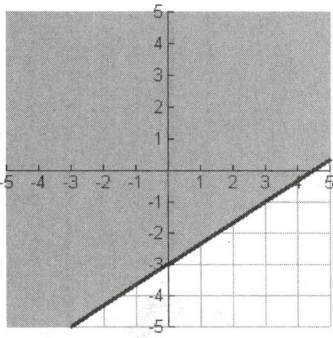

Name: Date:
Instructor: Section:

Practice Set 7.5

Match the inequality to all ordered pairs that are solutions of the inequality.

1. $y < -4$ A. $(0, 0)$ 1. _____

2. $x \geq 3$ B. $(-1, -2)$ 2. _____

3. $x - y < 3$ C. $(3, 1)$ 3. _____

4. $2x + 3y \leq 1$ D. $(-5, 3)$ 4. _____

                   E. No solutions listed

Graph each linear inequality. Label the axes accordingly.

5. $y > -4$

6. $x \leq -\dfrac{3}{2}$

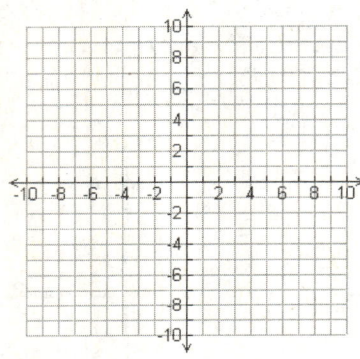

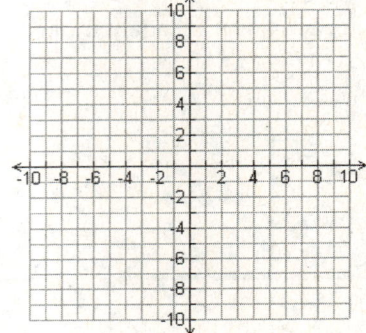

7. $y \geq \dfrac{7}{3}$

8. $x > 6$

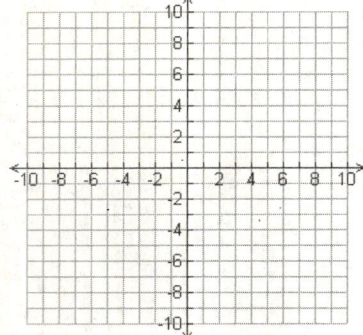

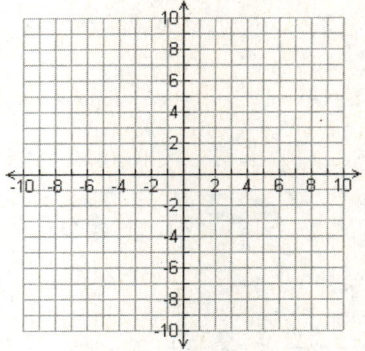

Practice Set 7.5

9. $y \leq 3x - 1$

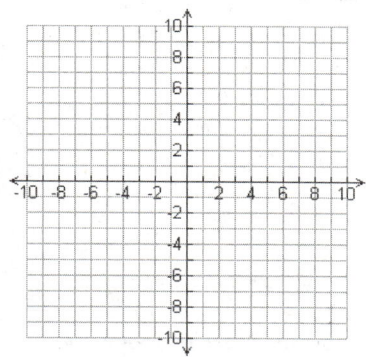

10. $y > \dfrac{4}{3}x + 2$

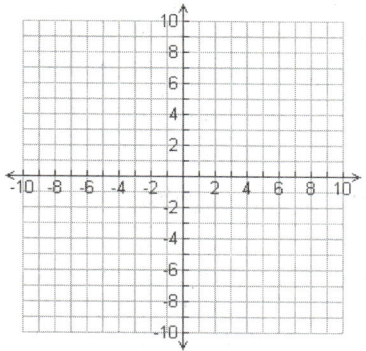

11. $x + 2y \geq 6$

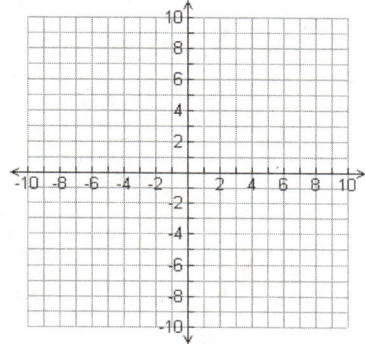

12. $2x - 3y \leq -9$

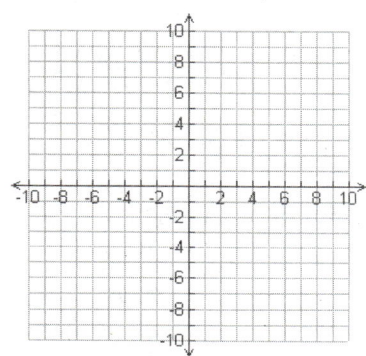

Insert either > or < in the blank so that the given point is a solution to the inequality.

13. $(-4, 8)$; $4x - 5y$ ____ 12

13. _____

14. $(5, 0)$; $x - y$ ____ -1

14. _____

15. $\left(-\dfrac{1}{4}, \dfrac{2}{3}\right)$; $8x + 15y$ ____ 9

15. _____

Insert either $<, >, \leq,$ or $\geq$ in the blank so that the given point is a solution to the inequality. More than one answer is possible.

16. $(2, -3)$; $3x + 6y$ ____ -12

16. _____

17. $\left(\dfrac{3}{5}, -\dfrac{4}{7}\right)$; $x + 4$ ____ 2

17. _____

Name:
Instructor:

Date:
Section:

7.6 Functions

> **Objectives**
> 1. Find the domain and range of a relation.
> 2. Recognize functions.
> 3. Evaluate functions.
> 4. Graph linear functions.
>
> **Key Vocabulary**
> relation, components of an ordered pair, domain, range, function, vertical line test, function notation $f(x)$, linear function

1 Find the domain and range of a relation.

Example 1 State the domain and range of the relation.

a) $\{(3,-1),(4,12),(-6,-1),(8,5),(7,3)\}$

b) $\{(1,-4),(7,2),(-1,-2),(-9,-3),(1,3)\}$

2 Recognize functions.

Example 2 Determine which of the following are functions.

a) $\{(3,-1),(4,12),(-6,-1),(8,5),(7,3)\}$

b) $\{(1,-4),(7,2),(-1,-2),(-9,-3),(1,3)\}$

c)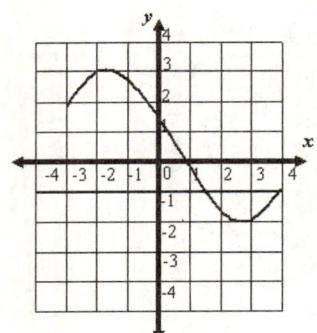

d)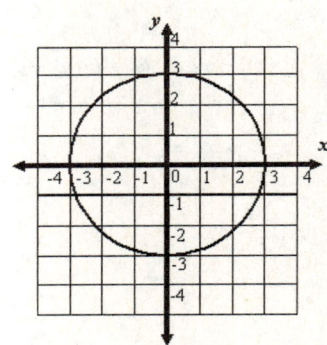

e)

domain   range

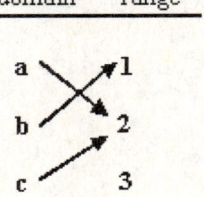

f)

domain   range

Answers: 1a) Domain: $\{-6,3,4,7,8\}$; Range: $\{-1,3,5,12\}$ b) Domain: $\{-9,-1,1,7\}$; Range: $\{-4,-3,-2,2,3\}$ 2a) yes   b) no   c) yes   d) no   e) no   f) yes

180   Angel/Runde, *Elementary Algebra for College Students*, 9e   Copyright © 2015 by Pearson Education, Inc.

Examples 7.6

3 Evaluate functions.

Example 3 Evaluate $f(x) = -2x + 1$ for the given values.

    a) $f(0)$ **b)** $f(3)$ **c)** $f(-4)$

Example 4 Evaluate $f(x) = 3x^2 - 2x + 1$ for the given values.

    a) $f(0)$ **b)** $f(2)$ **c)** $f(-5)$

4 Graph linear functions.

Example 5 Graph each linear function. Label the axes accordingly.

    a) $f(x) = 3x - 1$ **b)** $f(x) = -\dfrac{1}{2}x + 4$

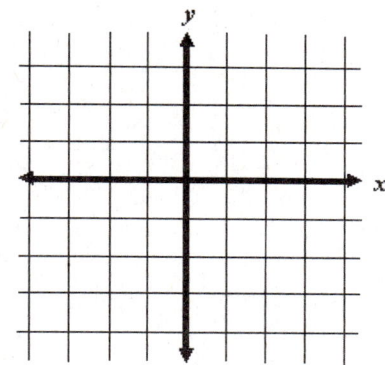
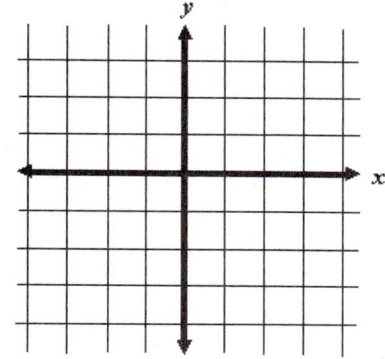

Answers: 3a) 1   b) −5   c) 9   4a) 1   b) 9   c) 86   5a)   b)

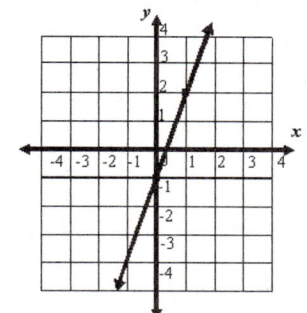
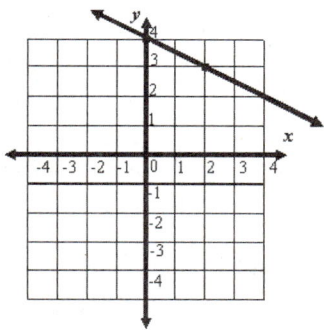

Name:
Instructor:
Date:
Section:

Practice Set 7.6

For each relation, **a)** state the domain and range, and **b)** determine whether the relation is a function.

1. $\{(1, -2), (3, 4), (-5, 6), (-8, 12), (54, 16)\}$

2. $\{(-3, 6), (4, 12), (-1, -8), (5, 6), (-2, -8)\}$

3. $\{(8, -2), (1, -5), (7, 0), (8, 2), (-1, 3)\}$

1. a)_____
   _____
   b)_____

2. a)_____
   _____
   b)_____

3. a)_____
   _____
   b)_____

Determine whether each relation is a function.

4.

Domain	Range
apple	yellow
eggplant	green
banana	red
grape	purple

5.

Domain	Range
apple	yellow
eggplant	green
banana	red
grape	purple

4. _____

5. _____

6.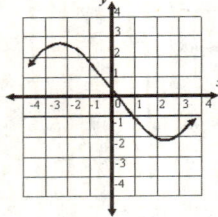

7.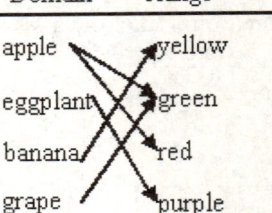

6. _____

7. _____

Evaluate each function for the given values.

8. $f(x) = -5x + 6$; $f(-1)$

9. $f(x) = \dfrac{3}{4}x - 8$; $f(-4)$

10. $f(x) = x^2 + x - 3$; $f(0)$

11. $f(x) = -4x^2 + 2x - 5$; $f(-2)$

12. $f(x) = \dfrac{2x - 1}{3}$; $f\left(\dfrac{1}{3}\right)$

8. _____

9. _____

10. _____

11. _____

12. _____

182   Angel/Runde, *Elementary Algebra for College Students*, 9e   Copyright © 2015 by Pearson Education, Inc.

Practice Set 7.6

Graph each linear function. Label the axes accordingly.

13. $f(x) = 4x - 5$

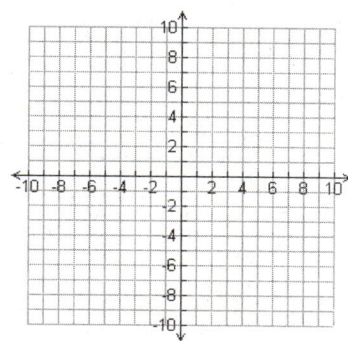

14. $f(x) = -x - 5$

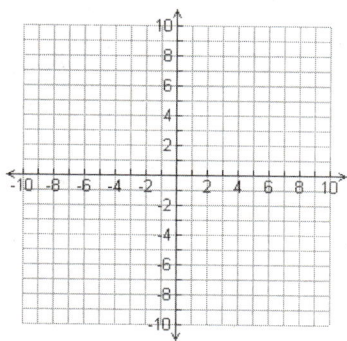

15. $f(x) = -3$

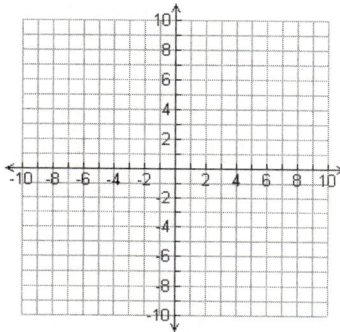

16. $f(x) = \frac{2}{3}x + 1$

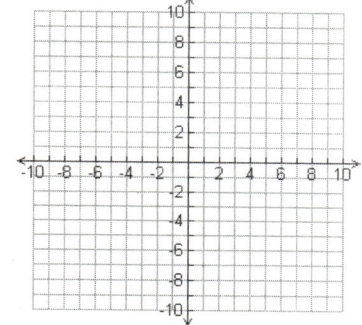

Problem Solving

17. An NFL player signs a contract that gives him a 4-million dollar signing bonus as well as 2 million dollars per year.
a) Write a function $f(x)$ for the total amount earned by the player after x years.
b) Sketch the graph of the function on the set of axes to the right (up to 5 years).
c) Use the graph to estimate the player's total earnings after 4 years.

17. a)

b)

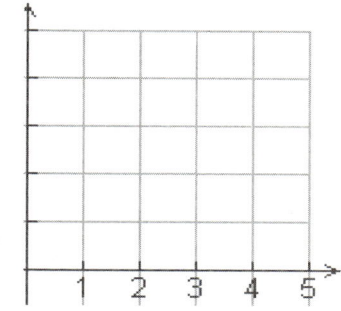

c)

Notes:

Chapter 7 Vocabulary Reference Sheet

Term	Definition	Example
graph	An illustration that shows the relationship between two variables in an equation	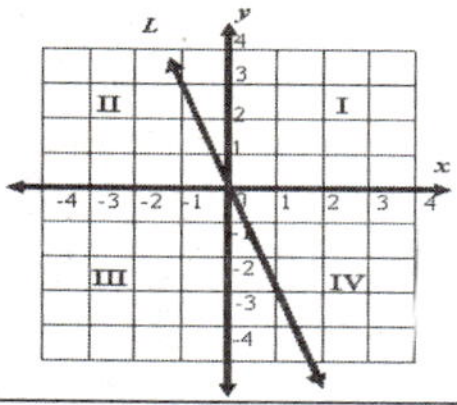
Cartesian (rectangular) coordinate system	A grid system for graphing points and equations	The line L above is graphed on the rectangular coordinate system.
collinear	A set of points that lie on a straight line	$(-1,3)$, $(0,2)$, and $(2,0)$ are collinear.
quadrants	The 4 sections of the rectangular coordinate system	See Roman numerals in graph above.
y-axis	The vertical axis of the rectangular coordinate system	See graph above.
x-axis	The horizontal axis of the rectangular coordinate system	See graph above.
origin	The point $(0,0)$ is the origin.	$(0,0)$
ordered pair	A pair of numbers used to represent a point (x,y) on a graph. The x-coordinate of the point is always written first.	$(-3,5)$ is an ordered pair.
coordinates	In (x,y), x and y are the coordinates.	In $(-3,5)$, -3 is the x-**coordinate** and 5 is the y-**coordinate**.
linear equation in two variables	An equation that can be put in the form $ax+by=c$, where a, b, and c are real numbers	$4x-3y=12$ $y=5x+3$ $x-3y+4=0$
standard form	$ax+by=c$ is the standard form of the equation of a line.	$2x+3y=6$ is the equation of a line in standard form.
x-intercept	The point at which a graph crosses the x-axis	$(-2,0)$
y-intercept	The point at which a graph crosses the y-axis	$(0,-2)$
horizontal line	The graph of an equation of the form $y=b$	$y=3$
vertical line	The graph of an equation of the form $x=a$	$x=-4$
slope	The ratio of the vertical change to the horizontal change between any two points on a graph	The slope of line L (above) is $m=\dfrac{-3}{1}=-3$.
parallel lines	Two lines that lie in the same plane but do not intersect. Parallel lines have the same slope and different y-intercepts	$y=-3x+5$ and $y=-3x-1$
perpendicular lines	Two lines that intersect to form right (90°) angles. The slopes of perpendicular lines are negative reciprocals. Any vertical line is perpendicular to any horizontal line.	$y=-3x+5$ and $y=\dfrac{1}{3}x-2$

Chapter 7 Vocabulary

negative reciprocals	Any two numbers whose product is −1	$m = -3$ and $m = \dfrac{1}{3}$ are negative reciprocals.
slope-intercept form	$y = mx + b$, where m is the slope and $(0, b)$ is the y-intercept of the line	$y = -3x + 5$ is the slope-intercept form for the line with $m = -3$ and $(0, 5)$.
point-slope form	$y - y_1 = m(x - x_1)$, where m is the slope of the line and (x_1, y_1) is a point on the line	$y - 2 = -3(x + 4)$ is the point-slope form for the line with $m = -3$ that goes through $(-4, 2)$.
linear inequality	An inequality of the form $ax + by < c$, $ax + by > c$, $ax + by \leq c$, or $ax + by \geq c$, where a, b, <u>and c are real numbers and a</u> and b are not both zero	$2x - 3y < 6$
relation	Any set of ordered pairs	$\{(-2, 0), (0, 3), (2, 5), (2, 8)\}$
domain	The set of first components in a relation	$\{(2, 3), (-2, 5)\}$ has domain $\{2, -2\}$.
range	The set of second components in a relation	$\{(2, 3), (-2, 5)\}$ has range $\{3, 5\}$.
function	A relation in which each first component corresponds to exactly one second component	$\{(-2, 0), (0, 3), (2, 5)\}$ but not $\{(-2, 0), (0, 3), (2, 5), (2, 8)\}$
function notation $f(x)$	Notation that is used when a function is represented by an equation To compute $f(a)$, substitute a for x.	$f(x) = -3x + 5$ $f(2) = -3(2) + 5 = -1$
vertical line test	If a vertical line intersects a graph in more than one point, the graph does not represent a function.	The line L (above) passes this test so it represents a function.
linear function	A function that can be written in the form $f(x) = ax + b$ The graph of a linear function is a non-vertical line.	$f(x) = -3x + 5$ or $f(x) = 5$, but not $x = 5$

Name:
Instructor:

Date:
Section:

Chapter 7 Practice Test A

Determine the quadrant in which the point lies or the axis on which it lies.

1. $(-2, 1)$

2. $(-4, -5)$

3. $(0, -3)$

4. $\left(5, \dfrac{7}{2}\right)$

1. _____

2. _____

3. _____

4. _____

Plot each point on the set of axes below. Label the point with its corresponding letter.

5. a) $(3, -2)$ b) $(-3, 2)$ c) $\left(\dfrac{7}{2}, 0\right)$

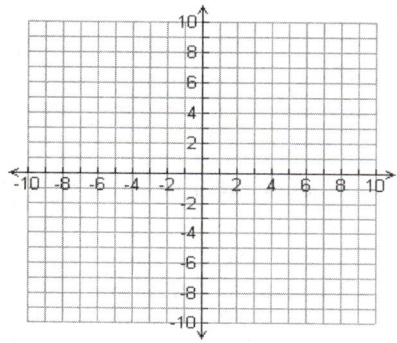

Determine which of the given points satisfy the equation.

6. $x - 3y = 8$; $(1, -2), (-1, -3), \left(0, -\dfrac{8}{3}\right)$

6. _____

Graph each linear equation by the method listed.

7. $x - 2y = -2$ by plotting points

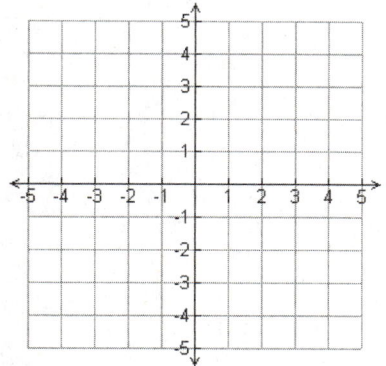

8. $6x + 9y = 18$ by using x- and y-intercepts

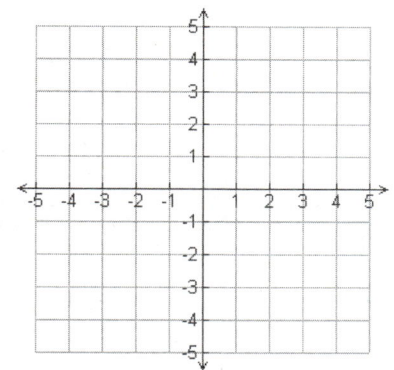

Practice Test 7A

9. $3x - 2y = 4$ by the slope-intercept method

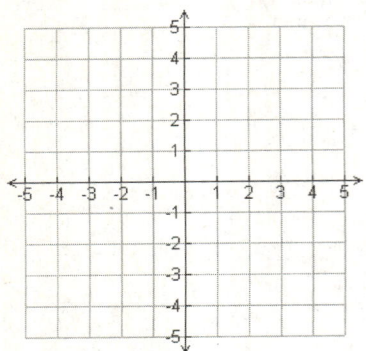

10. $x = 4$ by method of choice

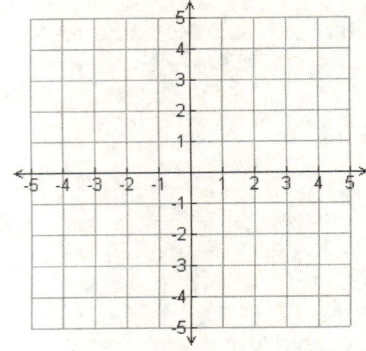

Determine the equation of each line L.

11.

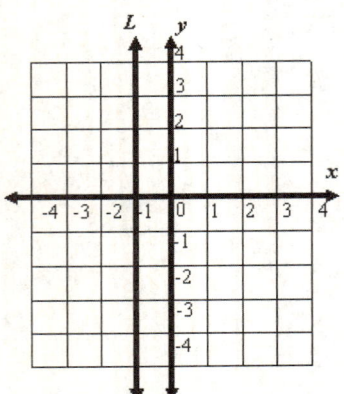

12.

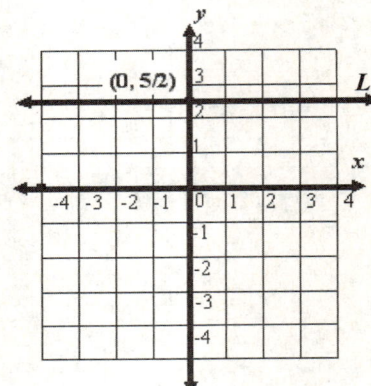

11. _____

12. _____

13. Determine the slope of the line that goes through the points $(-1, 5)$ and $(-3, 6)$.

13. _____

For each line L below, **a)** determine the slope and y-intercept, and **b)** write the equation of the line in slope-intercept form.

14.

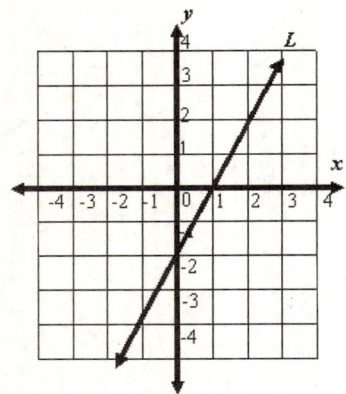

15.

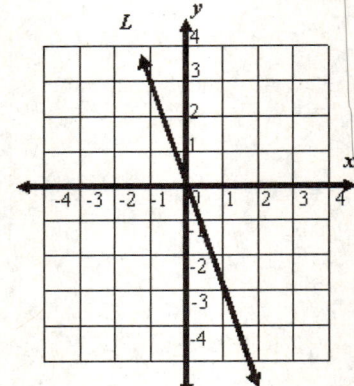

14. a) _____

_____

b) _____

15. a) _____

_____

b) _____

Practice Test 7A

16. Determine the slope of a line that is parallel to $3x + 5y = -10$.

16. _____

17. Determine the slope of a line that is perpendicular to $6x - y = 3$.

17. _____

18. Find the equation of the line that goes through the points $(8, -1)$ and $(4, 3)$. Write the equation in slope-intercept form.

18. _____

19. Find the equation of the vertical line that goes through the point $(-6, 5)$.

19. _____

Graph each linear inequality. Label the axes accordingly.

20. $3x + y \leq 5$ **21.** $2x - 3y > 0$

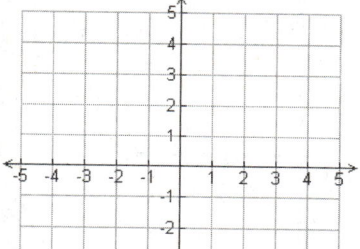
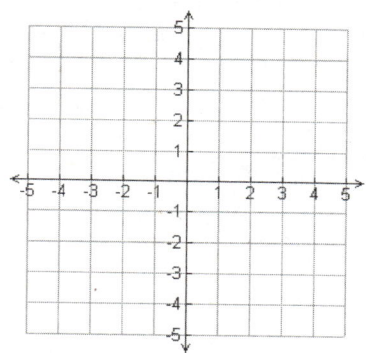

For the given relation, **a)** state the domain and range, and **b)** determine whether the relation is a function.

22. $\{(1, -2), (3, 4), (-5, 6), (-8, 12), (54, 16)\}$

22. a) _____

_____

b) _____

Determine whether the following statements are true (T) or false (F). If the statement is false, explain why.

23. Every function is a relation.

23. _____

_____

_____

24. All lines represent functions.

24. _____

_____

_____

Name:
Instructor:

Date:
Section:

Chapter 7 Practice Test B

1. The point $(-1, 4)$ lies in which quadrant?
   a) I b) II c) III d) IV

2. If an ordered pair has a positive x-coordinate and a negative y-coordinate, then the point lies in which quadrant?
   a) I b) II c) III d) IV

3. Which one of the following points satisfies the linear equation $x - 3y = -2$?
   a) $(0, 0)$ b) $(4, 2)$ c) $(-2, 1)$ d) $(4, -3)$

4. Which one of the following is the missing coordinate of the point $(-4, y)$ if the point is a solution of $3x - y = -5$?
   a) 7 b) 3 c) -7 d) -2

5. Which of the following represents the graph of $2x + 5y = -10$?

   a)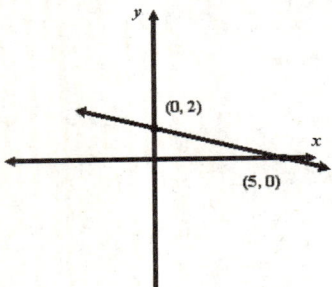
   b)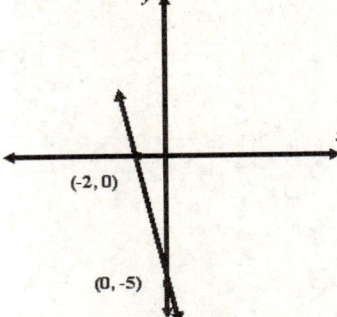

   c)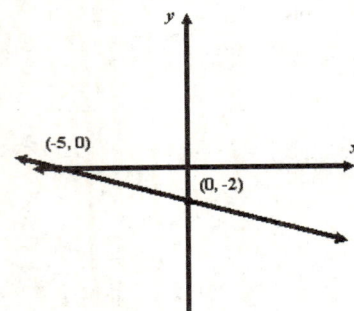
   d)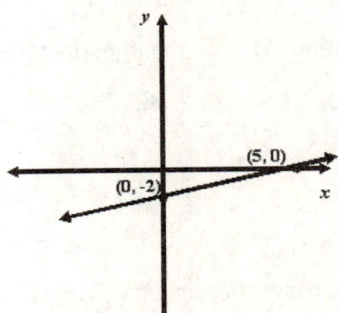

6. Find the slope through the points $(-2, -3)$ and $(-4, 1)$.
   a) -2 b) 2 c) $\frac{1}{2}$ d) $-\frac{1}{2}$

7. Which of the following represents the equation of a horizontal line that goes through the point $(3, -4)$?
   a) $y = 3x - 4$ b) $x = -4$ c) $x = 3$ d) $y = -4$

Practice Test 7B

8. Which of the following represents the equation for the line graphed below?

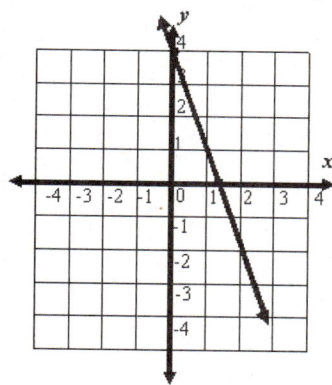

a) $y = -3x + 4$

b) $y = -\frac{1}{3}x - 4$

c) $y = 3x + 4$

d) $y = \frac{1}{3}x - 4$

9. Find the slope of the line graphed below.

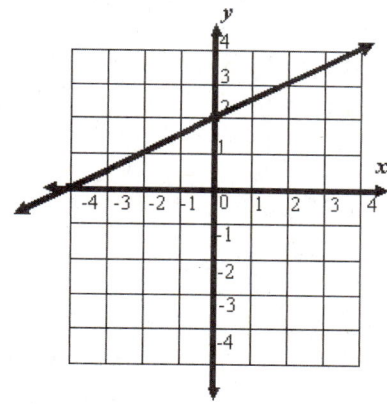

a) -2

b) 2

c) $\frac{1}{2}$

d) $-\frac{1}{2}$

10. Determine the slope of a line that is perpendicular to $x - 3y = 6$.

a) -3

b) 3

c) $\frac{1}{3}$

d) $-\frac{1}{3}$

11. Determine the x-intercept of $5x - 6y = 30$.

a) $(0, 6)$

b) $(-5, 0)$

c) $(6, 0)$

d) $(0, -5)$

12. Which of the following is the equation, in slope-intercept form, of the line that goes through the points $(3, -4)$ and $(-3, 4)$?

a) $y = \frac{4}{3}x$

b) $y = -\frac{4}{3}x$

c) $y = \frac{3}{4}x - 1$

d) $y = -\frac{4}{3}x - 1$

Practice Test 7B

13. Which one of the following points satisfies the linear inequality $5x - 3y < 8$?
    a) $(1, -1)$ b) $(4, -6)$ c) $(1, 2)$ d) $(5, 1)$

14. Which of the following represents the graph of $4x - 2y \geq 8$?

   a) 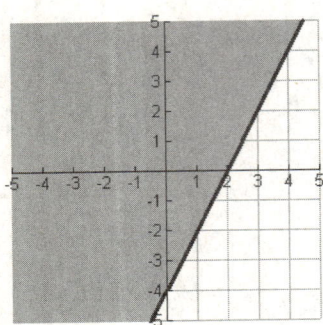   b)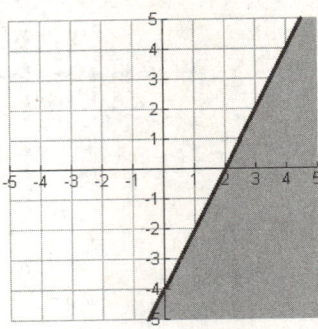

   c) 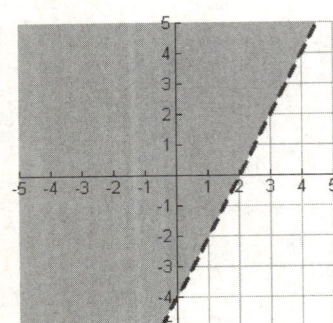   d)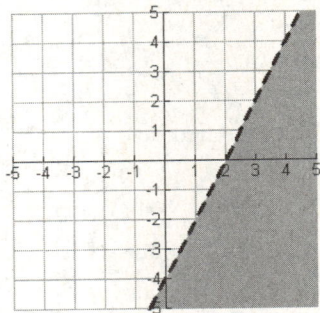

15. Let $f(x) = -3x - 1$. Find $f(2)$.
    a) 5   b) 7   c) -5 d) -7

16. Let $f(x) = 4x^2 - 2x + 8$. Find $f(-3)$.
    a) -22 b) -32 c) 50   d) 38

17. Which of the following represent functions?

   i) 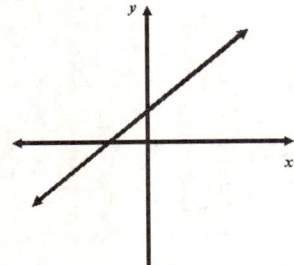   ii) 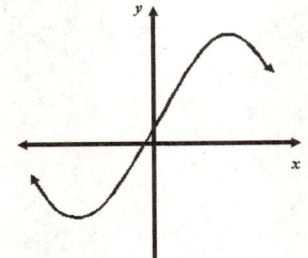   iii)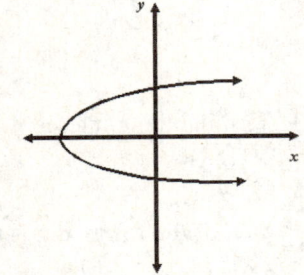

   a) i, ii and iii   b) i and ii only   c) ii and iii only   d) only i

Name:
Instructor:

Date:
Section:

8.1 Solving Systems of Equations Graphically

Objectives
1. Determine if an ordered pair is a solution to a system of equations.
2. Determine if a system of equations is consistent, inconsistent, or dependent.
3. Solve a system of equations graphically.

Key Vocabulary
system of linear equations, solution to a system of linear equations in two variables, consistent system of equations, inconsistent system of equations, dependent system of equations, solve a system of equations graphically

1 Determine if an ordered pair is a solution to a system of equations.

Example 1 Determine which of the following ordered pairs satisfy the system of equations.

$$y = 5x - 2$$
$$2x + 5y = -10$$

a) $(0, -2)$ b) $(1, 3)$ c) $\left(\dfrac{2}{5}, 0\right)$

2 Determine if a system of equations is consistent, inconsistent, or dependent.

Example 2 Determine whether the following system has exactly one solution, no solution, or an infinite number of solutions.

a) $\begin{array}{l} 6x - 2y = 4 \\ 3y = 9x - 6 \end{array}$ b) $\begin{array}{l} 2x + 4y = 4 \\ x + 6y = 6 \end{array}$ c) $\begin{array}{l} y = x + 1 \\ -x + y = 3 \end{array}$

Answers: 1a) yes b) no c) no   2a) infinite number of solutions b) one solution c) no solution

Practice Set 8.1

3 **Solve a system of equations graphically.**

Example 3 Solve the following systems of equations graphically.

a) $y = -2x + 5$
$-5x + 2y = 1$

b) $3x + y = 5$
$-x + 2y = -4$

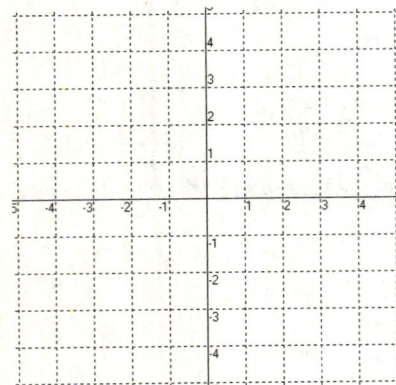

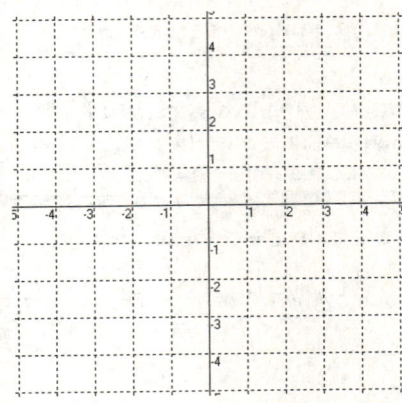

c) $y = \dfrac{1}{3}x + 6$
$-x + 3y = 18$

d) Sara likes to play video games. One arcade charges $0.25 per play while another arcade charges $1.00 plus $0.20 per play. We can represent this situation with the system of equations

$$c_1 = \dfrac{1}{4}x$$

$$c_2 = \dfrac{1}{5}x + 1$$

where c_1 is the total cost for the first arcade and c_2 is the total cost for the second arcade. How many video games, x, must Sara play so the total cost of each arcade is the same?

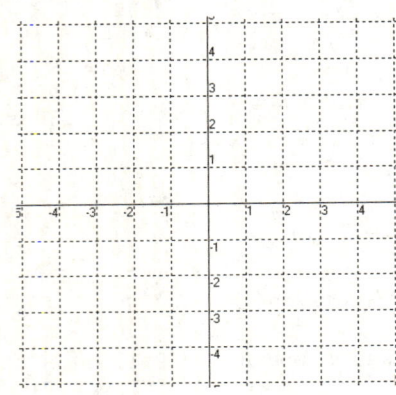

Answers: 3a) (1, 3) 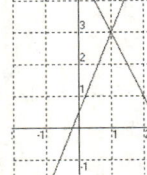 b) (2, −1) 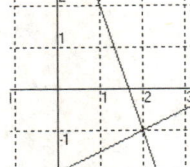 c) infinite number of solutions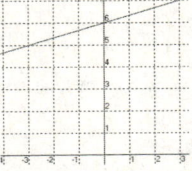

d) 20 games

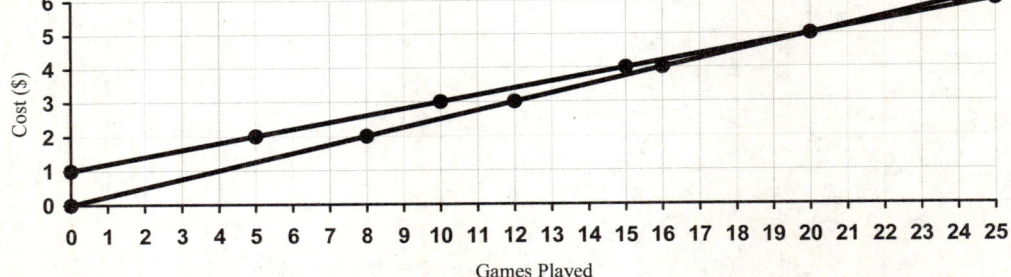

Practice Set 8.1

Determine which, if any, of the following ordered pairs satisfy each system of linear equations.

1. $x + y = 3$
    $x - y = 1$
        a) $(2, 1)$ b) $(1, 2)$ c) $(1, 0)$ 1._____

2. $5x - 3y = -2$
    $10x - y = 1$
        a) $(0, -1)$ b) $\left(0, \frac{2}{3}\right)$ c) $\left(\frac{1}{5}, 1\right)$ 2._____

Identify each system of linear equations (lines labeled 1 and 2) as consistent, inconsistent, or dependent. State whether the system has exactly one solution, no solution, or an infinite number of solutions.

3.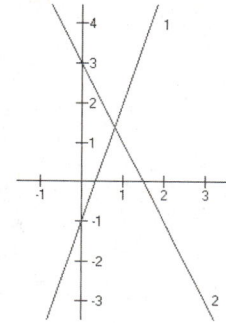

4.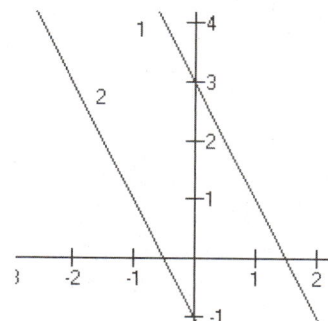

3._____

4._____

5.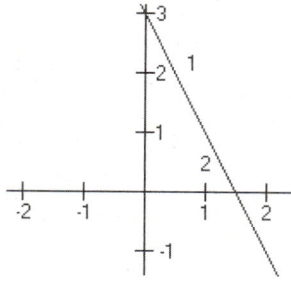

5._____

Write each equation in slope-intercept form. Without graphing the equations, state whether the system of equations has exactly one solution, no solution, or an infinite number of solutions.

6. $-2x + y = -1$
    $8x + 2y = 16$

7. $-3x - 9y = 7$
    $x + 3y = 12$

6._____

7._____

8. $x + y = 6$
    $-x + y = 1$

9. $-2x - 6y = 18$
    $x + 3y = -9$

8._____

9._____

Practice Set 8.1

Determine the solution to each system of equations graphically. If the system is dependent or inconsistent, so state.

10. $\quad y = 2x - 1$
$ y = -x + 11$

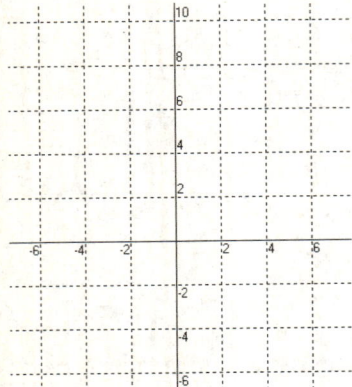

11. $\quad x - 2y = -4$
$ 2x - 4y = -8$

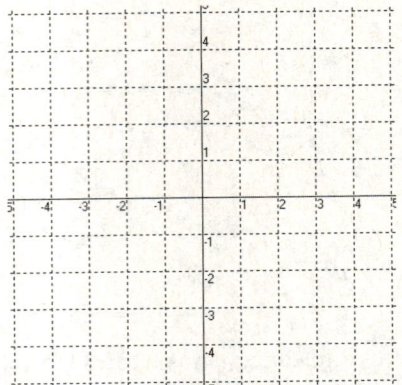

12. $\quad 7x + 6y = -9$
$ 2x + y = 1$

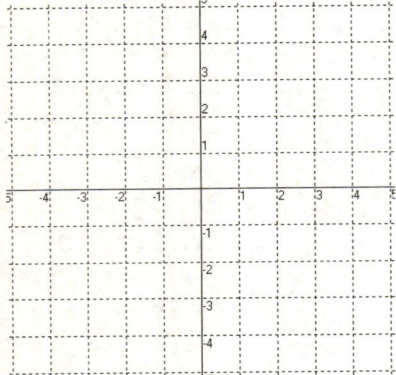

13. $\quad -x + 3y = -9$
$ -2x + 6y = 12$

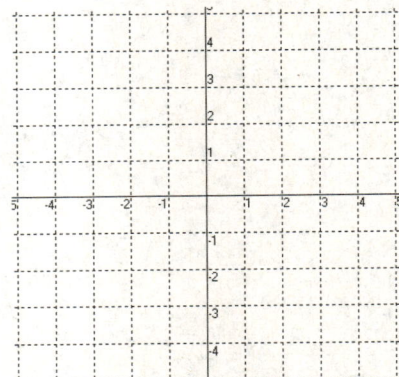

14. $\quad 2x - y = 1$
$ x - y = -1$

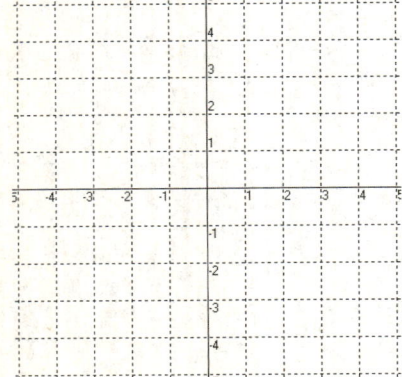

15. $\quad y = -2x + 9$
$ y = 3x - 1$

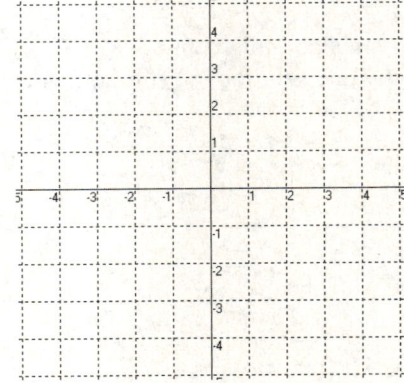

Name:
Instructor:
Date:
Section:

8.2 Solving Systems of Equations by Substitution

Objectives
1. Solve systems of equations by substitution.

Key Vocabulary
substitution method

1 Solve systems of equations by substitution.

Example 1 Find the solution of each system of equations by substitution.

a) $2x + y = 6$
$y = x + 3$

b) $2x + 3y = 5$
$2x + 4y = 8$

c) $2x + y = 3$
$-4x - 2y = 3$

d) $3x + y = 4$
$6x + 2y = 8$

e) $x + 2y = 4$
$3x - 2y = 6$

Example 2 Use a system of equations to solve each problem. Solve the system by substitution.

a) The sum of two positive integers is 50. Find the integers if one number is 12 greater than the other.

b) Eight people ate dinner for $80.50. The price for each adult was $12.50 and the price for each child was $6.00. How many adults and how many children attended the dinner?

Answers: 1a) (1, 4) b) (−2, 3) c) no solution d) infinite number of solutions e) $\left(\frac{5}{2}, \frac{3}{4}\right)$ 2a) 19 and 31 b) 5 adults and 3 children

Name:
Instructor:

Date:
Section:

Practice Set 8.2

Solve each system of equations by substitution.

1. $y = 2x - 1$
   $x + y = 5$

2. $x + y = 3$
   $x - y = 1$

3. $3x - 5y = 15$
   $-2x + y = 4$

4. $4x - 2y = 8$
   $6x - 3y = 12$

5. $4x - y = -1$
   $2x + 4y = 13$

6. $8x - 2y = 2$
   $4x - y = 2$

7. $5x - 3y = -3$
   $6x + 6y = -42$

8. $2x + 3y = 5$
   $y = 2x$

1._____

2._____

3._____

4._____

5._____

6._____

7._____

8._____

Use a system of equations to solve each problem.

9. The sum of two positive integers is 69. Find the integers if one number is 13 greater than the other.

9._____

10. Eight people ate dinner for $67.50. The price for each adult was $12.50 and the price for each child was $6.00. How many adults and how many children attended the dinner?

10._____

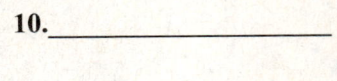

Name:
Instructor:
Date:
Section:

8.3 Solving Systems of Equations by the Addition Method

Objectives
1. Solve systems of equations by the addition method.

Key Vocabulary
addition (elimination) method

1 Solve systems of equations by the addition method.

Example 1 Solve each system of equations using the addition method.

a) $x + y = 4$
$x - y = 2$

b) $5x - 2y = 10$
$x - y = -1$

c) $7x - 6y = -1$
$2x - 4y = -2$

d) $-2x + 2y = 2$
$3x - 3y = 3$

e) $-4x + 3y = 6$
$12x + 9y = 18$

f) $-2x + 5y = -6$
$x - y = 0$

Example 2 Use a system of equations to solve each problem. Solve the system by the addition method.

a) One number is 2 more than twice another and their sum is 11. Find the numbers.

b) Carson has $1.75 in nickels and dimes in a jar. He has 8 more nickels than dimes. How many nickels and dimes does he have?

Answers: 1a) $(3, 1)$ b) $(4, 5)$ c) $\left(\dfrac{1}{2}, \dfrac{3}{4}\right)$ d) no solution e) $(0, 2)$ f) $(-2, -2)$ 2a) 3 and 8 b) 17 nickels and 9 dimes

Name:
Instructor:

Date:
Section:

Practice Set 8.3

Solve each system of equations using the addition method.

1. $x + y = 7$
   $-x + y = 3$

2. $2x + 3y = 3$
   $x - 3y = 6$

1._____

2._____

3. $x - 2y = -8$
   $2x + 5y = 11$

4. $4x - 2y = 8$
   $2x - y = 4$

3._____

4._____

5. $2x + 3y = 6$
   $2x + y = -2$

6. $2x - y = 2$
   $-2x + y = 2$

5._____

6._____

7. $5x + 3y = 5$
   $3x - 2y = 12$

8. $2x + 3y = 5$
   $3x - 4y = 2$

7._____

8._____

Use a system of equations to solve each problem.

9. One number is 1 more than twice another and their sum is 13. Find the numbers.

9._____

10. Carson has $2.15 in nickels and dimes in a jar. He has 10 more nickels than dimes. How many nickels and dimes does he have?

10._____

Name:
Instructor:

Date:
Section:

8.4 Systems of Equations: Applications and Problem Solving

Objectives
1. Use systems of equations to solve application problems.

Key Vocabulary
complementary angles, supplementary angles

1 Use systems of equations to solve application problems.

Example 1 Use a system of linear equations to find the solution. Use a calculator where appropriate.

a) The sum of two numbers is 6. If the second number is subtracted from the first, the result is −2. Find the numbers.

b) The perimeter of a rectangle is 36 feet. The width is 2 feet less than the length. Find the dimensions.

c) If a plane can travel 470 miles per hour with the wind and 410 miles per hour against the wind, find the speed of the plane in still air.

d) You invested $5000 in two accounts paying 2% and 9% annual interest, respectively. If the total interest earned for the year was $170, how much was invested at each rate?

e) Kennewick and Seattle are 192 miles apart. A car leaves Kennewick traveling towards Seattle, and another car leaves Seattle at the same time, traveling towards Kennewick. The car leaving Kennewick averages 10 mph more than the other, and they meet after 1 hour and 36 minutes. What are the average speeds of the cars?

f) A chemist has an 18% alcohol solution and a 9% alcohol solution. How many liters of each solution should the chemist mix together in order to get 360 liters of a 16% alcohol solution?

Answers: 1a) 2, 4   b) width 8 ft, length 10 ft   c) 440 mph   d) $4000 in the 2% account, $1000 in the 9% account
e) 65 mph, 55 mph   f) 80 L of 9% solution, 280 L of 18% solution

Name:
Instructor:

Date:
Section:

Practice Set 8.4

Use a system of linear equations to find the solution. Use a calculator where appropriate.

1. The sum of two numbers is 12. If the second number is subtracted from the first, the result is 3. Find the numbers.

   1._____

2. Angles A and B are complementary angles. If angle A is 2 degrees less than angle B, find the measure of each angle.

   2._____

3. The perimeter of a rectangle is 28 feet. The width is 2 feet less than the length. Find the dimensions.

   3._____

4. The population of City A is 50,000 and it is growing by 200 per year. The population of City B is 65,000 and it is decreasing by 550 per year. How long will it take for both cities to have the same population?

   4._____

5. If a plane can travel 520 miles per hour with the wind and 390 miles per hour against the wind, find the speed of the plane in still air.

   5._____

6. A man just purchased a high speed copier for his office and wants a service contract on the copier. He is considering two sources for the contract. Company A charges $25 a month plus 4 cents a copy. Company B charges $73 a month plus 2.5 cents a copy. How many copies would the man need to make per month for the monthly cost of both companies to be the same?

   6._____

Practice Set 8.4

7. You invested $4000 in two accounts paying 2% and 6% annual interest, respectively. If the total interest earned for the year was $120, how much was invested at each rate?

7._____

8. Lewiston and Pasco are 184 miles apart. A car leaves Lewiston traveling towards Pasco, and another car leaves Pasco at the same time traveling towards Lewiston. The car leaving Lewiston averages 20 miles per hour more than the other, and they meet after 1 hour and 36 minutes. What are the average speeds of the cars?

8._____

9. One canned juice drink is 30% orange juice; another is 5% orange juice. How many liters of each should be mixed together in order to get 25 liters that is 8% orange juice?

9._____

10. A woman is selecting tile for her foyer and living room. She wants to make a pattern using two different colors and types of tile. One type costs $3 per square foot tile and the other type costs $6 per square foot tile. She needs a total of 350 tiles but can only spend $1500 on tile. What is the maximum number of $6 tiles she can purchase?

10._____

Name:
Instructor:

Date:
Section:

8.5 Solving Systems of Linear Inequalities

Objectives
1. Solve systems of linear inequalities graphically.

Key Vocabulary
system of linear inequalities , solution to a system of linear inequalities

1 Solve systems of linear inequalities graphically.

Example 1 Determine the solution to each system of linear inequalities.

a) $3x - y > 6$
$x + y \leq 6$

b) $x + 2y < 4$
$x - y \leq 1$

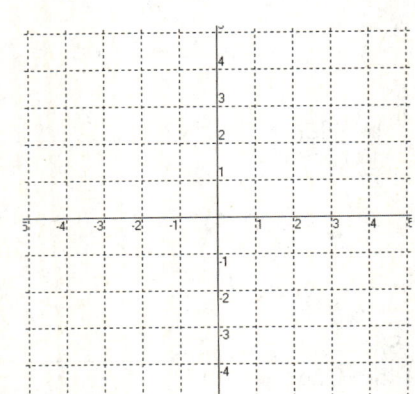
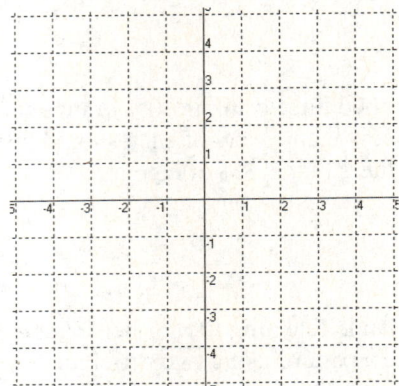

c) $x < 3$
$y \geq -2$

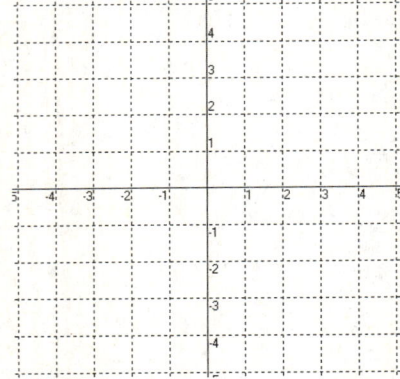

Answers: 1a)

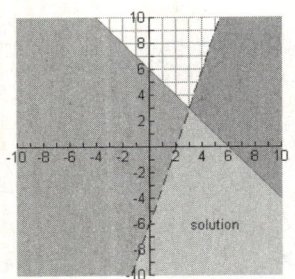

b)

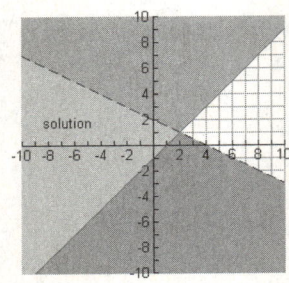

c)

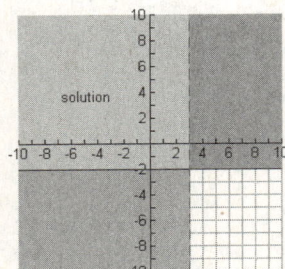

204   Angel/Runde, *Elementary Algebra for College Students*, 9e   Copyright © 2015 by Pearson Education, Inc.

Practice Set 8.5

Determine the solution to each system of linear inequalities.

1. $x + y > 2$
$x - y < 2$

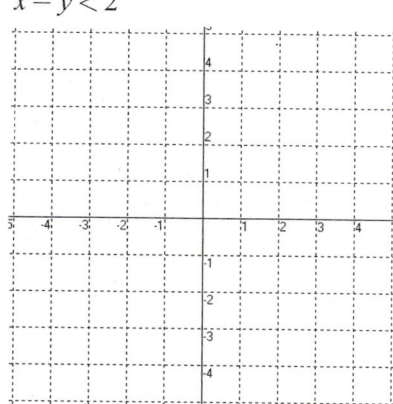

2. $y \leq x$
$y < -2x + 3$

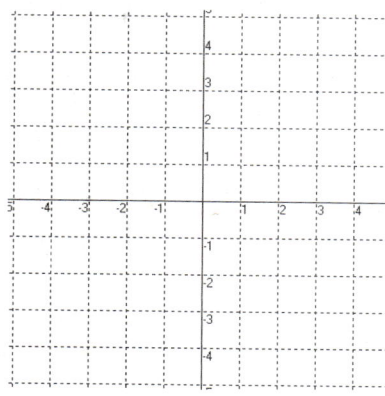

3. $y < -4x - 4$
$y \geq 2x + 2$

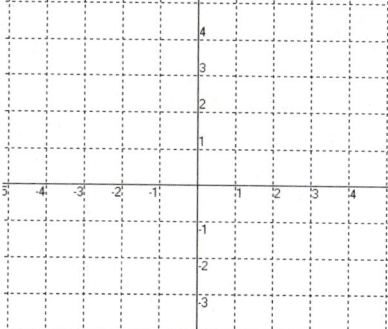

4. $x - 2y < 8$
$y \leq -x + 5$

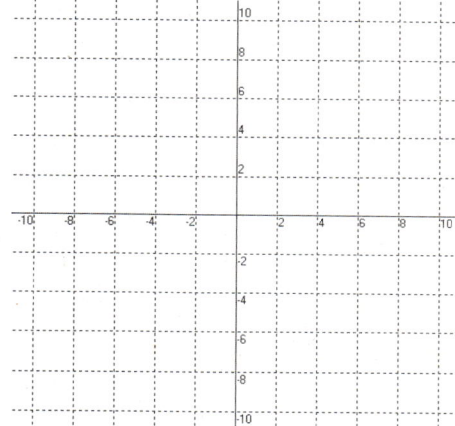

5. $x \leq 3$
$y \geq -8$

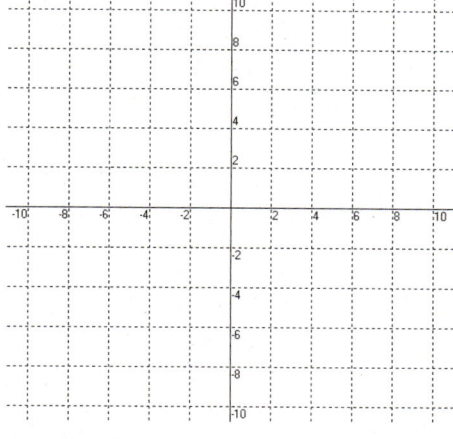

6. $x < 6$
$y < -2$

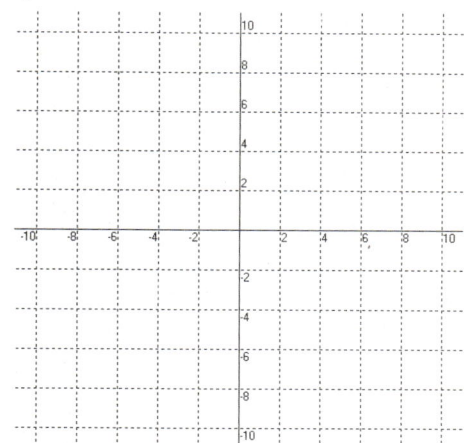

Notes:

Chapter 8 Vocabulary Reference Sheet

Term	Definition	Example
system of linear equations	A system having two or more linear equations	$x + y = 6$ $2x - y = 6$
solution to a system of linear equations	The ordered pair or pairs that satisfy all the equations in the system	(4, 2) is the solution of $\begin{array}{l} x+y=6 \\ 2x-y=6 \end{array}$ because it satisfies both equations in the system.
consistent system of equations	A system of equations that has exactly one solution	This system has exactly one solution, which is the ordered pair where the lines intersect.
inconsistent system of equations	A system of equations that has no solution	This system has no solution because the lines are parallel.
dependent system of equations	A system of equations that has an infinite number of solutions	This system has an infinite number of solutions because the graphs of both equations are the same line.
solving a system of equations graphically	Graphing the equations in a system on the same axes and determining where or whether the lines intersect	$x + y = 3$ $x - y = 1$ can be represented by the following graph, so the solution of this system is (2, 1).

Chapter 8 Vocabulary

substitution method	An algebraic method for solving a system of linear equations	$x + y = 3$ $y = x - 1$ Substitute $x - 1$ in for y and solve for x. $x + x - 1 = 3$ $2x = 4$ $x = 2$ Substitute 2 in for x and solve for y. $2 + y = 3$ $y = 1$ The solution is (2, 1).
addition (elimination) method	An algebraic method for solving a system of linear equations	$2x + y = 3$ $x - y = 9$ Add the equations and solve for x. $3x = 12$ $x = 4$ Substitute 4 in for x in one of the original equations and solve for y. $4 - y = 9$ $-y = 5$ $y = -5$ The solution is (4, −5).
complementary angles	Two angles are complementary angles if the sum of their measures is 90°.	If angle A is 32° and angle B is 58°, then angles A and B are complementary angles.
supplementary angles	Two angles are supplementary angles if the sum of their measures is 180°.	If angle A is 79° and angle B is 101°, then angles A and B are supplementary angles.
solution to a system of linear inequalities	The set of all points that satisfies all inequalities in the system	$3x - y > 6$ $x + y \leq 6$
solving a system of linear inequalities graphically	Graphing all linear inequalilties in a system on the same axes and determining the region where their graphs overlap	To solve a system of linear inequalities graphically, graph each inequality on the same axes (see **solution to a system of linear inequalities** above).

Name:
Instructor:
Date:
Section:

Chapter 8 Practice Test A

Determine which, if any, of the ordered pairs satisfy the system of equations.

1. $x + y = 4$
   $2x + 5y = 5$ (0, 4) $\left(1, \dfrac{3}{5}\right)$ (5, −1)

1. _____

Identify each system as consistent, inconsistent, or dependent. State whether the system has exactly one solution, no solution, or an infinite number of solutions.

2.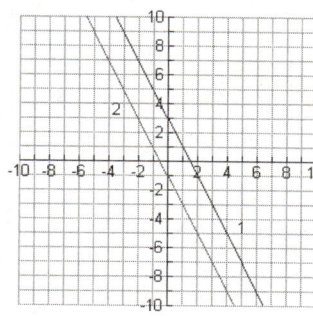

3.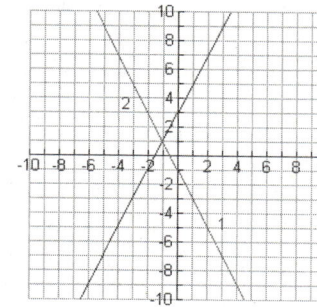

2. _____

3. _____

Write each equation in slope-intercept form. Then determine, without graphing or solving the system, whether the system of equations has exactly one solution, no solution, or an infinite number of solutions.

4. $3x - 3y = 5$
   $4x - 4y = 9$

5. $-5x + y = 4$
   $x + 3y = -4$

4. _____

5. _____

6. $2x + 4y = 6$
   $-x - 2y = -3$

6. _____

Practice Test 8A

Solve each system of equations graphically.

7. $3x - 5y = 7$
$y = 2x$

8. $-x + 2y = -6$
$y = \dfrac{1}{2}x + 1$

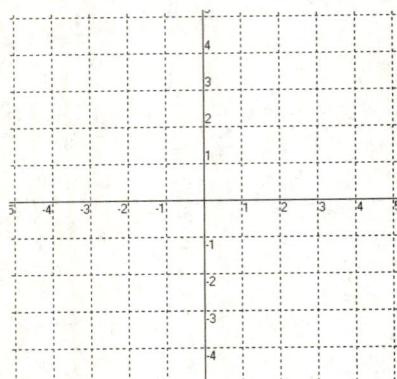
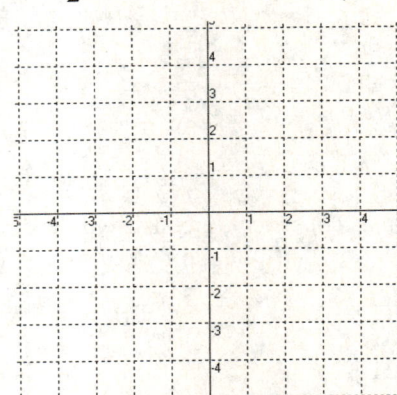

9. $3x - y = 10$
$4x + 3y = -4$

10. $-2x + 3y = -6$
$y = \dfrac{2}{3}x - 2$

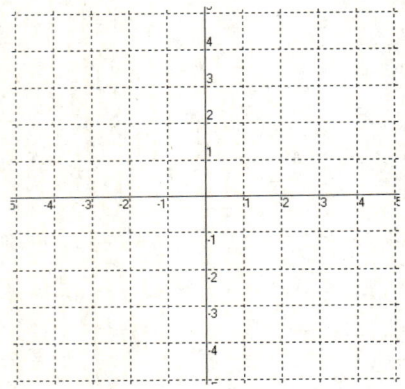
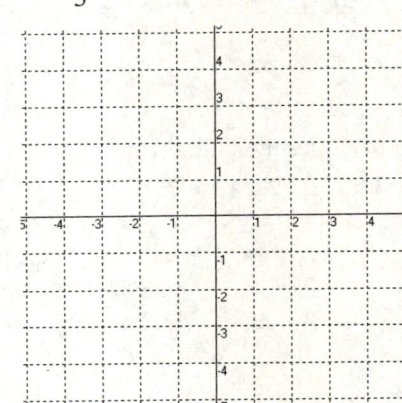

Solve each system of equations by substitution.

11. $x + y = -3$
$y = 3x + 1$

12. $3x - y = 2$
$x - y = 1$

11._____

12._____

13. $2x - 3y = 4$
$4x - 6y = 4$

13._____

Practice Test 8A

Solve each system of equations by using the addition method.

14. $\begin{array}{l} 2x - y = 7 \\ 3x + y = 3 \end{array}$

15. $\begin{array}{l} \frac{1}{3}x - y = 1 \\ x - 3y = 3 \end{array}$

14._____

15._____

16. $\begin{array}{l} 4x - 3y = 7 \\ -2x + 6y = -2 \end{array}$

16._____

Solve each system of equations using the method of your choice.

17. $\begin{array}{l} y = x + 3 \\ 2x - y = -1 \end{array}$

18. $\begin{array}{l} 6x - 4y = 10 \\ 3x + 2y = 1 \end{array}$

17._____

18._____

19. $\begin{array}{l} x + y = 1 \\ x - y = 2 \end{array}$

20. $\begin{array}{l} y = 2x + 7 \\ y = 2x + 6 \end{array}$

19._____

20._____

Use a system of linear equations to find the solution.

21. If a plane can travel 450 miles per hour with the wind and 400 miles per hour against the wind, find the speed of the plane in still air.

21._____

Practice Test 8A

22. You invested $10,000 in two accounts paying 2% and 5% annual interest, respectively. If the total interest earned for the year was $410, how much was invested at each rate?

22._____

23. A greens keeper at a local golf course wants to mix one type of grass seed that costs $4.30 per pound with a different grass seed that costs $1.80 per pound. How many pounds of each type of seed must be mixed to make a mixture of 200 pounds costing $2 per pound?

23._____

Determine the solution to each system of inequalities.

24. $-2x + y < 3$
$3x + y \geq -1$

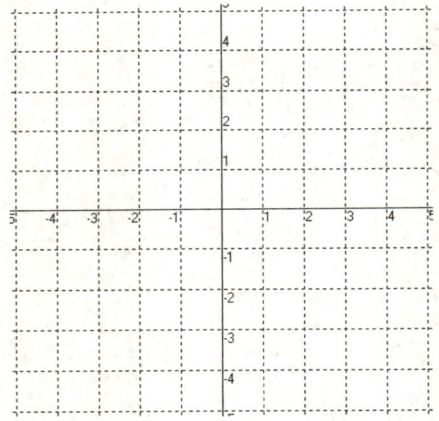

25. $-x + y > 2$
$y \leq 3$

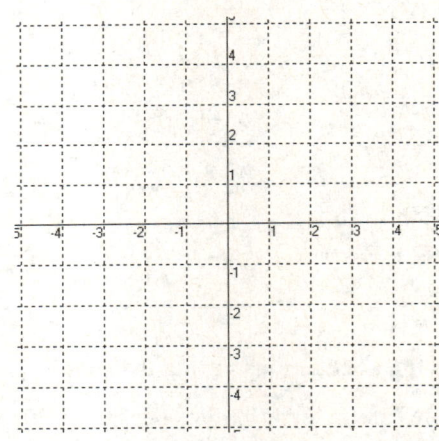

Name:
Instructor:
Date:
Section:

Chapter 8 Practice Test B

Determine which ordered pair satisfies the system of equations.

1. $2x + y = 2$
   $2x - y = 4$

   a) $(3, 2)$ b) $\left(1, \dfrac{3}{5}\right)$ c) $(1, 0)$ d) $\left(\dfrac{3}{2}, -1\right)$

Identify the system as consistent, inconsistent, or dependent. State whether the system has exactly one solution, no solution, or an infinite number of solutions.

2.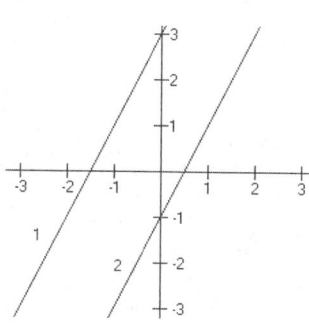

   a) inconsistent, no solution
   b) consistent, one solution
   c) dependent, infinite number of solutions
   d) consistent, no solution

3.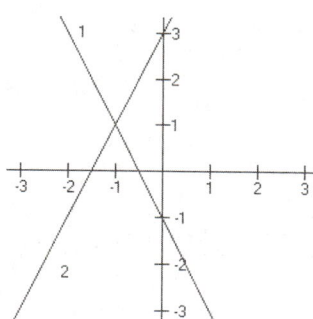

   a) inconsistent, no solution
   b) consistent, one solution
   c) dependent, infinite number of solutions
   d) consistent, no solution

Write each equation in slope-intercept form. Then determine, without graphing or solving the system, whether the system of equations has exactly one solution, no solution, or an infinite number of solutions.

4. $x - 5y = 5$
   $x + 5y = -5$

   a) exactly one solution b) no solution c) infinite number of solutions

5. $2x - 5y = 7$
   $-4x + 10y = -7$

   a) exactly one solution b) no solution c) infinite number of solutions

6. $3x - 4y = 5$
   $5x - 2y = 9$

   a) exactly one solution b) no solution c) infinite number of solutions

Practice Test 8B

Choose the graph that represents each system of equations.

a) b)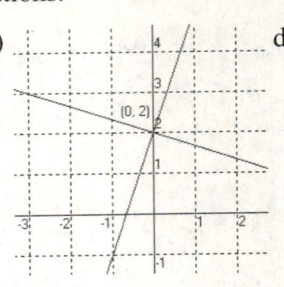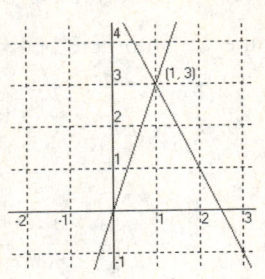

7. $3x + y = 0$
   $2x - y = -5$

   a) b) c) d)

8. $3x - y = -2$
   $x + 3y = 6$

   a) b) c) d)

9. $3x - y = 0$
   $3x - y = -5$

   a) b) c) d)

10. $y = 3x$
    $y = -2x + 5$

    a) b) c) d)

Solve each system of equations by substitution.

11. $y = x + 5$
    $x + 2y = 4$

    a) $(1, 6)$ b) $(-2, 3)$ c) $(0, 2)$ d) $(8, -2)$

12. $x - 2y = -3$
    $2x - y = 6$

    a) $(-3, 0)$ b) $(3, 0)$ c) $(1, 2)$ d) $(5, 4)$

13. $2x - y = -4$
    $2x + 2y = -4$

    a) $(-2, 0)$ b) $(-1, 2)$ c) $(0, -2)$ d) $(2, 2)$

Solve each system of equations by using the addition method.

14. $3x + 5y = 4$
    $-x + y = 4$

    a) $(0, 4)$ b) $(1, 5)$ c) $(-2, 2)$ d) $(3, -1)$

Practice Test 8B

15. $\dfrac{2}{3}x - 2y = 2$
    $x - 3y = 2$

    a) no solution
    b) infinite number of solutions
    c) $(2, 0)$
    d) $(0, -1)$

16. $2x - 4y = 6$
    $4x + 2y = 27$

    a) $\left(1, \dfrac{23}{2}\right)$
    b) $(1, -1)$
    c) $\left(6, \dfrac{3}{2}\right)$
    d) $\left(2, -\dfrac{1}{2}\right)$

Solve each system of equations using the method of your choice.

17. $3x - 2y = 6$
    $2x - y = 3$

    a) $(3, 0)$
    b) $(0, -3)$
    c) $\left(1, -\dfrac{3}{2}\right)$
    d) $(1, -1)$

18. $x - y = 5$
    $-x + y = -5$

    a) $(1, -4)$
    b) $(-2, -7)$
    c) no solution
    d) infinite number of solutions

19. $3x + 4y = 1$
    $x - 7y = -3$

    a) $\left(1, \dfrac{4}{7}\right)$
    b) $\left(-\dfrac{7}{3}, 2\right)$
    c) $\left(0, \dfrac{1}{4}\right)$
    d) $\left(-\dfrac{1}{5}, \dfrac{2}{5}\right)$

20. $-x + 3y = 6$
    $x = -3$

    a) $(-3, 3)$
    b) $(-3, -3)$
    c) $(-3, 1)$
    d) $(-3, -1)$

Use a system of linear equations to find the solution.

21. If a plane can travel 500 miles per hour with the wind and 400 miles per hour against the wind, find the speed of the plane in still air.

    a) 450 mph
    b) 480 mph
    c) 520 mph
    d) 430 mph

22. You invested $20,000 in two accounts paying 12% and 5% annual interest, respectively. If the total interest earned for the year was $1868, how much was invested at each rate?

    a) $15,000 at 12%, $5000 at 5%
    b) $12,400 at 12%, $7600 at 5%
    c) $13,500 at 12%, $6500 at 5%
    d) $12,000 at 12%, $8000 at 5%

Practice Test 8B

23. A gardener wants to mix two types of fertilizer, one type costs $5 per pound and the other type costs $3 per pound. How many pounds of each type of fertilizer must be mixed to make a mixture of 200 pounds costing $4.50 per pound?

   a) 100 pounds of $5, 100 pounds of $3
   b) 75 pounds of $5, 125 pounds of $3
   c) 150 pounds of $5, 50 pounds of $3
   d) 115 pounds of $5, 85 pounds of $3

Graph each system of inequalities.

24. $x - 6y > 6$
    $-x + y \leq 4$

a)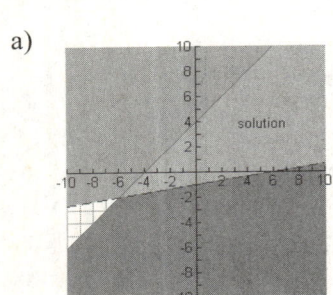

b)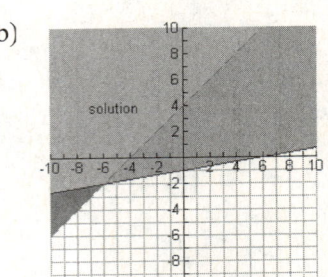

c)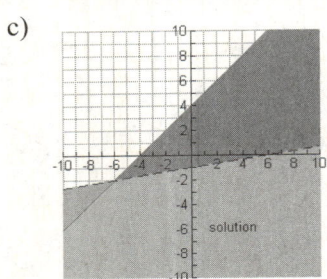

d)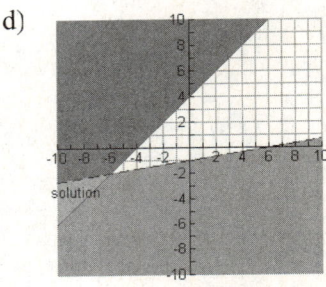

25. $y \leq 3$
    $x + 2y > 8$

a)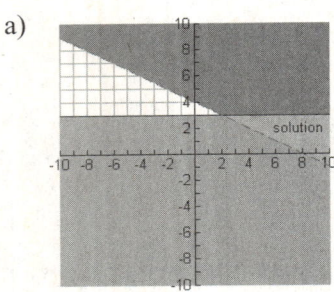

b)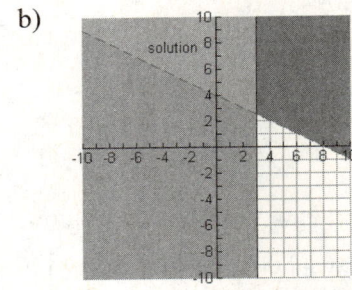

c)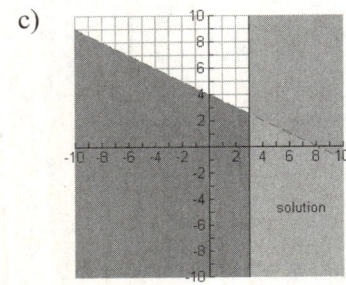

d)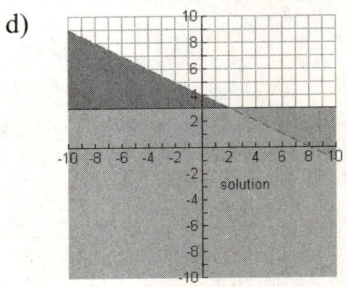

Name:
Instructor:

Date:
Section:

9.1 Evaluating Square Roots

Objectives
1. Evaluate square roots of real numbers.
2. Recognize that not all square roots represent real numbers.
3. Determine whether the square root of a real number is rational or irrational.
4. Write square roots as exponential expressions.

Key Vocabulary
square root, principal (positive) square root, negative square root, radical sign, radicand, radical expression, index, imaginary numbers, perfect squares, perfect square factor, rational numbers, irrational numbers

1 Evaluate square roots of real numbers.

Example 1 Evaluate.

a) $\sqrt{64}$ 	b) $\sqrt{169}$ 	c) $\sqrt{\dfrac{9}{25}}$

d) $-\sqrt{36}$ 	e) $-\sqrt{144}$ 	f) $-\sqrt{\dfrac{4}{25}}$

2 Recognize that not all square roots represent real numbers.

Example 2 Indicate whether the radical expression is a real or imaginary number.

a) $-\sqrt{4}$ 	b) $\sqrt{-4}$

3 Determine whether the square root of a real number is rational or irrational.

Example 3 Use your calculator or **Table 9.1** in your textbook to determine whether the following square roots are rational or irrational numbers.

a) $\sqrt{225}$ 	b) $\sqrt{17}$ 	c) $\sqrt{39}$ 	d) $\sqrt{121}$

4 Write square roots as exponential expressions.

Example 4 Write each radical expression in exponential form.

a) $\sqrt{8}$ 	b) $\sqrt{11}$ 	c) $\sqrt{6x}$ 	d) $\sqrt{16x^2y}$

Answers: 1a) 8 b) 13 c) $\dfrac{3}{5}$ d) –6 e) –12 f) $-\dfrac{2}{5}$ 2a) real b) imaginary 3a) rational b) irrational c) irrational d) rational

4a) $8^{1/2}$ b) $11^{1/2}$ c) $(6x)^{1/2}$ d) $(16x^2y)^{1/2}$

Name:
Instructor:

Date:
Section:

Practice Set 9.1

Evaluate each square root.

1. $-\sqrt{169}$
2. $\sqrt{9}$

3. $\sqrt{121}$
4. $-\sqrt{81}$

5. $-\sqrt{25}$
6. $\sqrt{16}$

7. $-\sqrt{\dfrac{9}{64}}$
8. $\sqrt{\dfrac{100}{169}}$

Write in exponential form.

9. $\sqrt{21}$
10. $\sqrt{15xy}$

11. $\sqrt{18x^2}$
12. $\sqrt{39x^3y^2}$

Classify each number as rational, irrational, or imaginary.

13. $\sqrt{27}$
14. 52.671

1. _____

2. _____

3. _____

4. _____

5. _____

6. _____

7. _____

8. _____

9. _____

10. _____

11. _____

12. _____

13. _____

14. _____

Practice Set 9.1

15. $\sqrt{-25}$ **16.** $-\sqrt{\dfrac{5}{18}}$

15._____

16._____

Indicate whether each statement is true or false.

17. $\sqrt{5}$ is a real number.   **18.** $\sqrt{400}$ is an integer.

17._____

18._____

19. $\sqrt{25}$ is an irrational number.   **20.** $\sqrt{52}$ is a rational number.

19._____

20._____

21. $-\sqrt{19}$ is imaginary.   **22.** $\sqrt{-25}$ is a real number.

21._____

22._____

23. Arrange the following list from smallest to largest. Do not use a calculator or **Table 9.1**.

$\sqrt{4}, \dfrac{1}{2}, -\sqrt{16}, 5, \sqrt{7}, 3.5$

23._____

Name:
Instructor:

Date:
Section:

9.2 Simplifying Square Roots

Objectives
1. Use the product rule to simplify square roots containing constants.
2. Use the product rule to simplify square roots containing variables.

Key Vocabulary
product rule for square roots

1 Use the product rule to simplify square roots containing constants.

Example 1 Simplify.

a) $\sqrt{28}$ b) $\sqrt{75}$ c) $\sqrt{84}$ d) $\sqrt{240}$

2 Use the product rule to simplify square roots containing variables.

Example 2 Simplify. Assume all variables represent positive numbers.

a) $\sqrt{x^{24}}$ b) $\sqrt{a^6 b^{16}}$ c) $\sqrt{x^5}$ d) $\sqrt{a^{15}}$

Example 3 Simplify. Assume all variables represent positive numbers.

a) $\sqrt{25x^7}$ b) $\sqrt{90y^2}$ c) $\sqrt{120x^6 y^5}$ d) $\sqrt{315x^5 y^{11}}$

Example 4 Multiply and then simplify. Assume all variables represent positive numbers.

a) $\sqrt{12} \cdot \sqrt{3}$ b) $\left(\sqrt{7x}\right)^2$ c) $\sqrt{3xy^3} \cdot \sqrt{9x^2 y^2}$ d) $\sqrt{11a^3 b} \cdot \sqrt{6ab}$

Answers: 1a) $2\sqrt{7}$ b) $5\sqrt{3}$ c) $2\sqrt{21}$ d) $4\sqrt{15}$ 2a) x^{12} b) $a^3 b^8$ c) $x^2\sqrt{x}$ d) $a^7\sqrt{a}$ 3a) $5x^3\sqrt{x}$ b) $3y\sqrt{10}$ c) $2x^3 y^2\sqrt{30y}$ d) $3x^2 y^5\sqrt{35xy}$ 4a) 6 b) $7x$ c) $3xy^2\sqrt{3xy}$ d) $a^2 b\sqrt{66}$

Name:
Instructor:
Date:
Section:

Practice Set 9.2

Simplify. Assume all variables represent positive numbers.

1. $\sqrt{27}$ 2. $\sqrt{80}$ 3. $\sqrt{175}$

1._____

2._____

3._____

4. $\sqrt{54}$ 5. $\sqrt{480}$ 6. $\sqrt{112}$

4._____

5._____

6._____

7. $\sqrt{ab^2c^3}$ 8. $\sqrt{x^4y^3z^{10}}$ 9. $\sqrt{m^7n^{11}}$

7._____

8._____

9._____

10. $\sqrt{7x^4y}$ 11. $\sqrt{224a^2b^6}$ 12. $\sqrt{135xy^8z^9}$

10._____

11._____

12._____

Practice Set 9.2

Simplify. Assume all variables represent positive numbers.

13. $\sqrt{13} \cdot \sqrt{13}$ 14. $\sqrt{48} \cdot \sqrt{5}$ 13._____

14._____

15. $\sqrt{27} \cdot \sqrt{12}$ 16. $\sqrt{3x} \cdot \sqrt{6x}$ 15._____

16._____

17. $\sqrt{5x^4y^3} \cdot \sqrt{27x^2y^3}$ 18. $\sqrt{8a^7} \cdot \sqrt{16a^2b^8}$ 17._____

18._____

19. $\left(\sqrt{3x}\right)^2$ 20. $\left(\sqrt{15x^4y^2}\right)^2$ 19._____

20._____

21. $\sqrt{25x^2}\sqrt{2xy^2}$ 22. $\left(\sqrt{5ab}\right)^2\left(\sqrt{15ab}\right)^2$ 21._____

22._____

Name:
Instructor:
Date:
Section:

9.3 Adding, Subtracting, and Multiplying Square Roots

Objectives
1. Add and subtract square roots.
2. Multiply square roots.

Key Vocabulary
like radicals, like square roots, unlike radicals, unlike square roots

1 Add and subtract square roots.

Example 1 Simplify if possible.

a) $5\sqrt{2} + 3\sqrt{2} - 5$

b) $6\sqrt{3} - \sqrt{x} + 2\sqrt{3} + 2\sqrt{x}$

c) $5\sqrt{2} + 6\sqrt{3}$

d) $x + 4\sqrt{x} + 9\sqrt{x} + 7$

e) $a\sqrt{a} + 4\sqrt{a} - 5a$

f) $\sqrt{8} + \sqrt{2}$

g) $\sqrt{20} - \sqrt{45}$

h) $3\sqrt{18} + \sqrt{8}$

i) $\sqrt{32} - 2\sqrt{45} + 3 + 5\sqrt{125}$

2 Multiply square roots.

Example 2 Multiply. Assume all variables represent positive numbers.

a) $\sqrt{5}(\sqrt{15} - 3)$

b) $\sqrt{3x}(\sqrt{2x} + \sqrt{3y})$

Example 3 Multiply using the FOIL method. Assume all variables represent positive numbers.

a) $(2 - \sqrt{7})(5 + 3\sqrt{7})$

b) $(\sqrt{5} + \sqrt{11})(\sqrt{5} - \sqrt{11})$

c) $(a - \sqrt{2b})(a + \sqrt{2b})$

Answers: 1a) $8\sqrt{2} - 5$ b) $8\sqrt{3} + \sqrt{x}$ c) cannot be simplified d) $x + 13\sqrt{x} + 7$ e) $(a+4)\sqrt{a} - 5a$ f) $3\sqrt{2}$ g) $-\sqrt{5}$ h) $11\sqrt{2}$ i) $4\sqrt{2} + 19\sqrt{5} + 3$ 2a) $5\sqrt{3} - 3\sqrt{5}$ b) $x\sqrt{6} + 3\sqrt{xy}$ 3a) $-11 + \sqrt{7}$ b) -6 c) $a^2 - 2b$

Name:
Instructor:

Date:
Section:

Practice Set 9.3

Simplify each expression.

1. $5\sqrt{6} - \sqrt{6}$

2. $4\sqrt{7} + 5\sqrt{7} - 3$

3. $5\sqrt{x} - 3\sqrt{x} + \sqrt{x} + 5$

4. $2\sqrt{3} + 3\sqrt{a} - 7\sqrt{3} + 2\sqrt{a}$

5. $\sqrt{54} + \sqrt{24}$

6. $\sqrt{125} - \sqrt{80}$

7. $2\sqrt{8} - 3\sqrt{18}$

8. $5\sqrt{40} + 2\sqrt{90}$

Multiply. Assume all variables represent positive numbers.

9. $5(\sqrt{a} - \sqrt{10})$

10. $\sqrt{x}(\sqrt{x} + \sqrt{xy})$

11. $(6 + \sqrt{3})(5 - \sqrt{3})$

12. $(\sqrt{5} - \sqrt{8y})(2\sqrt{5} - \sqrt{2y})$

13. $(6 + \sqrt{5})(6 - \sqrt{5})$

14. $(\sqrt{3} - 2\sqrt{2x})(\sqrt{3} + 2\sqrt{2x})$

15. Find the perimeter and area of a square with sides of length $\sqrt{5} - 2$.

1. _____

2. _____

3. _____

4. _____

5. _____

6. _____

7. _____

8. _____

9. _____

10. _____

11. _____

12. _____

13. _____

14. _____

15. _____

_____

Name:
Instructor:
Date:
Section:

9.4 Dividing Square Roots

Objectives
1. Understand what it means for a square root to be simplified.
2. Use the quotient rule to simplify square roots.
3. Rationalize denominators.
4. Rationalize denominators using conjugates.

Key Vocabulary
a simplified square root, quotient rule for square roots, rationalize a denominator, conjugates

1 Understand what it means for a square root to be simplified.

A square root is simplified when
1. No radicand has a factor that is a perfect square.
2. No radicand contains a fraction.
3. No denominator contains a square root.

Example 1 Explain why each square root is not fully simplified.

a) $\dfrac{15}{\sqrt{7}}$ b) $6\sqrt{\dfrac{3}{5}}$ c) $\sqrt{2y^4}$

2 Use the quotient rule to simplify square roots.

Example 2 Simplify. Assume all variables represent positive numbers.

a) $\sqrt{\dfrac{75}{3}}$ b) $\dfrac{\sqrt{50}}{\sqrt{2}}$ c) $\sqrt{\dfrac{200x^3y^5}{18x^5y^3}}$ d) $\dfrac{\sqrt{384m^4n^5}}{\sqrt{12mn}}$

3 Rationalize denominators.

Example 3 Simplify. Assume all variables represent positive numbers.

a) $\dfrac{1}{\sqrt{7}}$ b) $\dfrac{5}{\sqrt{12}}$ c) $\sqrt{\dfrac{17}{3}}$ d) $\sqrt{\dfrac{a^2}{80}}$

4 Rationalize denominators using conjugates.

Example 4 Simplify. Assume all variables represent positive numbers.

a) $\dfrac{12}{\sqrt{3}-1}$ b) $\dfrac{22}{\sqrt{6}+\sqrt{17}}$ c) $\dfrac{3\sqrt{x}}{13-\sqrt{x}}$ d) $\dfrac{3}{\sqrt{a}+b}$

Answers: 1a) square root in denominator b) fraction in radicand c) radicand has perfect square factor 2a) 5 b) 5 c) $\dfrac{10y}{3x}$ d) $4mn^2\sqrt{2m}$ 3a) $\dfrac{\sqrt{7}}{7}$ b) $\dfrac{5\sqrt{3}}{6}$ c) $\dfrac{\sqrt{51}}{3}$ d) $\dfrac{a\sqrt{5}}{20}$ 4a) $6\sqrt{3}+6$ b) $-2\sqrt{6}+2\sqrt{17}$ c) $\dfrac{-39\sqrt{x}-3x}{x-169}$ d) $\dfrac{3\sqrt{a}-3b}{a-b^2}$

Name:
Instructor:

Date:
Section:

Practice Set 9.4

Simplify each expression. Assume all variables represent positive numbers.

1. $\sqrt{\dfrac{80}{5}}$

2. $\dfrac{\sqrt{63}}{\sqrt{7}}$

3. $\sqrt{\dfrac{25}{169}}$

4. $\sqrt{\dfrac{36x^2 y}{6x}}$

5. $\sqrt{\dfrac{7a^2 b^5}{28ab^2 z}}$

6. $\dfrac{\sqrt{54x^2 y^6}}{\sqrt{3x^4 z^4}}$

7. $\sqrt{\dfrac{3}{5}}$

8. $\sqrt{\dfrac{4}{24}}$

9. $\dfrac{\sqrt{x}}{\sqrt{y}}$

10. $\sqrt{\dfrac{x^5}{20}}$

11. $\dfrac{\sqrt{160x^4 y}}{\sqrt{2x^5 y^5}}$

12. $\dfrac{7}{\sqrt{2}+5}$

13. $\dfrac{5}{\sqrt{7}-\sqrt{2}}$

14. $\dfrac{\sqrt{5}}{\sqrt{a}+\sqrt{5}}$

15. $\dfrac{9\sqrt{x}}{3-\sqrt{x}}$

1. _____

2. _____

3. _____

4. _____

5. _____

6. _____

7. _____

8. _____

9. _____

10. _____

11. _____

12. _____

13. _____

14. _____

15. _____

Name:
Instructor:

Date:
Section:

9.5 Solving Radical Equations

> **Objectives**
> 1. Solve radical equations containing only one square root.
> 2. Solve radical equations containing two square roots.
>
> **Key Vocabulary**
> radical equation, isolating the square root, extraneous root

1 Solve radical equations containing only one square root.

Example 1 Solve each equation.

a) $\sqrt{x} = 5$

b) $\sqrt{x-5} = 7$

c) $\sqrt{x} = -25$

d) $4\sqrt{x} = x + 3$

e) $\sqrt{y} + 6 = 15$

f) $3x + 2\sqrt{x} - 1 = 0$

2 Solve radical equations containing two square roots.

Example 2 Solve each equation. If the equation has no real solution, so state.

a) $\sqrt{3x+6} = \sqrt{4x-3}$

b) $3\sqrt{x-3} - \sqrt{6x+12} = 0$

Answers: 1a) 25   b) 54   c) not a real number   d) 1, 9   e) 81   f) $\frac{1}{9}$ 2a) 9   b) 13

Name:
Instructor:

Date:
Section:

Practice Set 9.5

Solve each equation. If the equation has no real solution, so state.

1. $\sqrt{x} = 4$
2. $\sqrt{x-8} = 5$

3. $\sqrt{y-5} = 8$
4. $\sqrt{5x+21} - 3 = x$

5. $\sqrt{x+9} = 5$
6. $\sqrt{4r-7} = \sqrt{3r+1}$

7. $\sqrt{10x-7} = 3\sqrt{x}$
8. $2\sqrt{x} = \sqrt{x+12}$

9. $6\sqrt{x} = x+8$
10. $\sqrt{5x-9} = \sqrt{2x-3}$

11. $\sqrt{2x-1} = x-2$
12. $\sqrt{3y+5} = \sqrt{2y+11}$

1. _____

2. _____

3. _____

4. _____

5. _____

6. _____

7. _____

8. _____

9. _____

10. _____

11. _____

12. _____

13. Find the value of the indicated variable when the area of the rectangle is known.

$A = 15$, width $\sqrt{x+4}$, length 5

13. _____

Name:
Instructor:
Date:
Section:

9.6 Radicals: Applications and Problem Solving

Objectives
1. Use the Pythagorean Theorem.
2. Use the distance formula.
3. Use radicals to solve application problems.

Key Vocabulary
Pythagorean Theorem, distance formula

1 Use the Pythagorean Theorem.

Example 1

Use the Pythagorean Theorem to find the missing leg of a right triangle whose hypotenuse is 7 inches and other leg is $\sqrt{3}$ inch.

Example 2

A rectangular field is 130 meters long from end to end. Find the length of the diagonal, to the nearest hundredth, from one end to the opposite corner of the other if the width is 58.1 meters.

2 Use the distance formula.

Example 3 Use the distance formula to find the length of the line segment between the given points. Give the exact value, then round your answer to the nearest hundredth.

a) $(-2, 8)$ and $(1, 5)$

b) $(-1, -1)$ and $(2, 5)$

3 Use radicals to solve application problems.

Example 4

The period of a pendulum depends on the length of the string, Scott made a pendulum with a string measuring 34 inches. What is the period for this pendulum? (Use $T = 2\pi\sqrt{\dfrac{L}{32}}$, where T is the period in seconds and L is the length of the pendulum in feet.)

Answers: 1. $\sqrt{46}$ in. 2. 142.39 m 3a) $3\sqrt{2} \approx 4.24$ b) $3\sqrt{5} \approx 6.71$ 4. $2\pi\sqrt{\dfrac{34/12}{32}} \approx 1.87$ sec

Name:
Instructor:

Date:
Section:

Practice Set 9.6

Use the Pythagorean Theorem to find each indicated quantity. Give the exact value, then round your answer to the nearest hundredth.

1.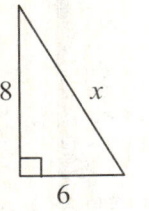

2.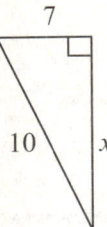

1._____

2._____

3.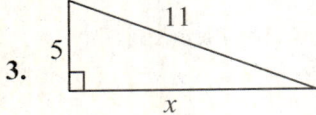

3._____

4. **Ladder on a House** Braden needs to wash a window on his house that is 15 feet above the ground. The closest the bottom of the ladder can be to the house is 8 feet because of a garden. How long does the ladder need to be in order for Braden to reach the window?

4._____

5. Find the distance between the points (6, −1) and (2, 2).

5._____

6. Find the distance between the points (4, −1) and (0, −5).

6._____

7. The period of a simple pendulum depends on the length of the string. Scott made a pendulum with a string measuring 24 inches. What is the period for this pendulum? (Use $T = 2\pi\sqrt{\dfrac{L}{32}}$, where T is the period in seconds and L is the length of the pendulum in feet.)

7._____

8. A cylindrical bowl can hold water 3 inches deep. If the volume of the bowl is 600 cubic inches, find the radius of the bowl. Use the formula $V = \pi r^2 h$.

8._____

Name:
Instructor:

Date:
Section:

9.7 Higher Roots and Rational Exponents

Objectives
1. Evaluate cube and fourth roots.
2. Simplify cube and fourth roots.
3. Write radical expressions in exponential form.

Key Vocabulary
cube root of *a*, fourth root of *a*, perfect cube, perfect fourth powers, product rule for radicals, exponential form of a radical expression, negative exponent rule

1 Evaluate cube and fourth roots.

Example 1 Evaluate.

a) $\sqrt[3]{125}$

b) $\sqrt[3]{-64}$

2 Simplify cube and fourth roots.

Example 2 Simplify.

a) $\sqrt[3]{120}$

b) $\sqrt[4]{96}$

3 Write radical expressions in exponential form.

Example 3 Write each radical in exponential form, then simplify.

a) $\sqrt[4]{x^8}$

b) $\sqrt[3]{y^{27}}$

c) $\sqrt[3]{a^3}$

Example 4 Simplify and write the answer in radical form.

a) $\sqrt[12]{x^3}$

b) $\sqrt[15]{y^3}$

Example 5 Evaluate.

a) $27^{4/3}$

b) $81^{3/4}$

c) $64^{-4/3}$

Example 6 Simplify.

a) $\sqrt[4]{49^2}$

b) $\sqrt[6]{64^2}$

c) $\sqrt{a} \cdot \sqrt[3]{a}$

d) $\left(\sqrt[3]{x^2}\right)^6$

Answers: 1a) 5 b) −4   2a) $2\sqrt[3]{15}$ b) $2\sqrt[4]{6}$ 3a) x^2 b) y^9 c) a 4a) $\sqrt[4]{x}$ b) $\sqrt[5]{y}$ 5a) 81 b) 27 c) $\dfrac{1}{256}$

6a) 7 b) 4 c) $a^{5/6}$ or $\sqrt[6]{a^5}$ d) x^4

Name:
Instructor:

Date:
Section:

Practice Set 9.7

Evaluate.

1. $\sqrt[3]{-64}$ 2. $\sqrt[3]{216}$ 3. $\sqrt[4]{256}$

1._____

2._____

3._____

Simplify.

4. $\sqrt[3]{24}$ 5. $\sqrt[4]{405}$ 6. $\sqrt[5]{192}$

4._____

5._____

6._____

Write each radical in exponential form and then simplify.

7. $\sqrt[3]{x^5}$ 8. $\sqrt[4]{a^{12}}$ 9. $\sqrt[5]{y^5}$

7._____

8._____

9._____

10. $\sqrt[4]{25^2}$ 11. $\sqrt[6]{8^2}$ 12. $\sqrt[5]{3^{15}}$

10._____

11._____

12._____

Write each radical in exponential form and then simplify. Write the answer in simplified radical form.

13. $\sqrt[6]{x^2}$ 14. $\sqrt[8]{y^4}$ 15. $\sqrt[15]{a^3}$

13._____

14._____

15._____

Write each expression in radical form and then evaluate.

16. $125^{2/3}$ 17. $36^{3/2}$ 18. $8^{-2/3}$

16._____

17._____

18._____

Write each radical in exponential form and then simplify. Write the answer in exponential form.

19. $\sqrt[3]{x} \cdot \sqrt[5]{x}$ 20. $\left(\sqrt[3]{x^5}\right)^3$

19._____

20._____

Chapter 9 Vocabulary Reference Sheet

Term	Definition	Example
square roots	The numbers that, when squared, result in the given number	The square roots of 25 are 5 and -5 since $5^2 = 25$ and $(-5)^2 = 25$.
principal square root	The positive square root of a number a is written as $\sqrt{a}$	$\sqrt{36} = 6$
negative square root	The negative square root of a number a is written as $-\sqrt{a}$.	$-\sqrt{36} = -6$
radical sign	The symbol $\sqrt{}$	
radicand	The number or expression inside the radical sign	The number 7 in the expression $\sqrt{7}$
radical expression	The entire expression, including the radical sign and the radicand	The entire expression $\sqrt{2x^2y}$
index	The index tells the "root" of the expression	The number 3 in the expression $\sqrt[3]{8}$
perfect squares	The numbers that are squares of natural numbers	$1, 4, 9, 16, 25, 36, \ldots$
rational numbers	Numbers that can be written in the form $\dfrac{a}{b}$, where a and b are integers, and $b \neq 0$. When a rational number is written as a decimal, it will be either a terminating or repeating decimal.	$\dfrac{2}{3}, -5, 0, 2.625, 1.\overline{3}$
irrational numbers	Numbers that are not rational numbers. Irrational numbers when written as decimals are nonterminating, nonrepeating decimals.	$\pi, 1.325419\ldots, \sqrt{2}$
product rule for square roots	The product of two square roots is equal to the square root of the product $\sqrt{a} \cdot \sqrt{b} = \sqrt{a \cdot b}$, provided $a \geq 0, b \geq 0$.	$\sqrt{2} \cdot \sqrt{8} = \sqrt{2 \cdot 8} = \sqrt{16} = 4$
like square roots	Square roots having the same radicand	$-\sqrt{7}$ and $3\sqrt{7}$
like radicals	Radicals having the same index and the same radicand	$-\sqrt{7}$ and $\sqrt{11}$
unlike square roots	Square roots having different radicands	$5\sqrt{2}$ and $2\sqrt{3}$
unlike radicals	Radicals having different indices or different radicands	$\sqrt{10}$ and $\sqrt[3]{10}$
a simplified square root	A square root in which no radicand has a factor that is a perfect square, no radicand contains a fraction, and no denominator contains a square root	$\sqrt{8}, \sqrt{\dfrac{1}{2}}, \dfrac{2}{\sqrt{9}}$ and not simplified but $\sqrt{6}$ is simplified.
quotient rule for square roots	The quotient of two square roots is equal to the square root of the quotient $\sqrt{\dfrac{a}{b}} = \dfrac{\sqrt{a}}{\sqrt{b}}$, provided $a \geq 0, b > 0$.	$\sqrt{\dfrac{4}{25}} = \dfrac{\sqrt{4}}{\sqrt{25}} = \dfrac{2}{5}$

Chapter 9 Vocabulary

rationalize a denominator	To rewrite a radical expression containing a radical in the denominator with an equivalent one that does not contain a radical in the denominator	$\dfrac{1}{\sqrt{2}} = \dfrac{1}{\sqrt{2}} \cdot \dfrac{\sqrt{2}}{\sqrt{2}} = \dfrac{\sqrt{2}}{2}$
conjugates	A binomial having the same two terms with the sign of the second term changed	The conjugate of $2\sqrt{x}+3$ is $2\sqrt{x}-3$.
radical equation	An equation that contains a variable in a radicand	$\sqrt{x+2} = 4$
extraneous root	A number obtained when solving an equation that is not a solution to the original equation	The number 16 is an extraneous root to the equation $\sqrt{x} = -4$.
Pythagorean Theorem	The square of the hypotenuse of a right triangle is equal to the sum of the squares of the two legs. $(\text{leg})^2 + (\text{leg})^2 = (\text{hypotenuse})^2$	Use the Pythagorean Theorem to find the hypotenuse of a right triangle whose legs are 3 and 4. $3^2 + 4^2 = x^2$ $25 = x^2$ $5 = x = $ hypotenuse
distance formula	$d = \sqrt{(x_2 - x_1)^2 + (y_2 - y_1)^2}$	The distance between the points $(-5, 4)$ and $(3, -2)$ is $d = \sqrt{(-5-3)^2 + (4-(-2))^2}$ $d = \sqrt{(-8)^2 + (6)^2}$ $d = \sqrt{64 + 36}$ $d = \sqrt{100} = 10$
cube root of a	$\sqrt[3]{a} = b$ if $b^3 = a$	The cube root of 8 is 2 since $2^3 = 8$.
fourth root of a	$\sqrt[4]{a} = b$ if $b^4 = a$, $a \geq 0$	The fourth root of 81 is 3 since $3^4 = 81$.
product rule for radicals	$\sqrt[n]{a} \cdot \sqrt[n]{b} = \sqrt[n]{ab}$, for $a \geq 0$, $b \geq 0$ and n is even $\sqrt[n]{a} \cdot \sqrt[n]{b} = \sqrt[n]{ab}$, when n is odd	$\sqrt[4]{4} \cdot \sqrt[4]{4} = \sqrt[4]{4 \cdot 4} = \sqrt[4]{16} = 2$ $\sqrt[3]{4} \cdot \sqrt[3]{16} = \sqrt[3]{4 \cdot 16} = \sqrt[3]{64} = 4$
exponential form of a radical expression	$\sqrt[n]{a} = a^{\frac{1}{n}}$, for $a \geq 0$ and n is even $\sqrt[n]{a} = a^{\frac{1}{n}}$, when n is odd $\sqrt[n]{a^m} = \left(\sqrt[n]{a}\right)^m = a^{\frac{m}{n}}$, for $a \geq 0$ and m and n integers	$\sqrt{16} = 16^{\frac{1}{2}} = 4$ $\sqrt[3]{27} = 27^{\frac{1}{3}} = 3$ $\sqrt[4]{16^3} = \left(\sqrt[4]{16}\right)^3 = 16^{\frac{3}{4}} = 8$
negative exponent rule	$a^{-m} = \dfrac{1}{a^m}$	$3^{-2} = \dfrac{1}{3^2} = \dfrac{1}{9}$

Name:
Instructor:

Date:
Section:

Chapter 9 Practice Test A

1. Write $\sqrt{7y}$ in exponential form.

    1._____

2. Write $y^{3/5}$ in radical form.

    2._____

Simplify. Assume all variables represent positive numbers.

3. $\sqrt{64}$ 4. $\sqrt{60}$

    3._____

    4._____

5. $\sqrt{54a^2}$ 6. $\sqrt{96x^5 y^4}$

    5._____

    6._____

7. $\sqrt{9xy^4}\sqrt{45xy}$ 8. $\sqrt{6x^2 y^3}\sqrt{3xy^3}$

    7._____

    8._____

9. $\sqrt{\dfrac{6}{216}}$ 10. $\dfrac{\sqrt{5a^2 d^3}}{\sqrt{5a^4 d}}$

    9._____

    10._____

11. $\dfrac{7}{\sqrt{11}}$ 12. $\sqrt{\dfrac{25x}{7}}$

    11._____

    12._____

13. $\sqrt{\dfrac{135xy^2}{5x^2 y^6}}$ 14. $\dfrac{12}{3+\sqrt{x}}$

    13._____

    14._____

Practice Test 9A

15. $\dfrac{5}{\sqrt{5}-2}$ 16. $2\sqrt{24}+3\sqrt{54}-\sqrt{6}$ 15. _____

16. _____

17. $\sqrt{x}-5\sqrt{x}+4\sqrt{x}$ 17. _____

Solve.

18. $\sqrt{2x-1}=3$ 19. $\sqrt{6x+7}-2=x$ 18. _____

19. _____

20. Find the length indicated by x. Give the exact value, then round your answer to the nearest hundredth.   20. _____

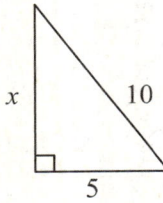

21. Find the length of the line segment between the points $(5, 3)$ and $(-3, -1)$. Give the exact value, then round your answer to the nearest hundredth.   21. _____

22. Evaluate $16^{-3/2}$.   22. _____

23. Simplify $\sqrt[3]{x^4}\cdot\sqrt[3]{x^2}$.   23. _____

24. An object was dropped from 88 feet. At what velocity did the object hit the ground? Use the formula $v=\sqrt{2gh}$ with $g=32$ feet per second squared. The velocity will be in feet per second.   24. _____

25. If a square window has an area of 64 square feet, find the length of one side of the window.   25. _____

Name: Date:
Instructor: Section:

Chapter 9 Practice Test B

1. Write $\sqrt[3]{9x}$ in exponential form.

   a) $9x^{1/3}$ b) $(9x)^{1/3}$ c) $9x^{2/3}$ d) $(9x^{1/3})$

2. Write $x^{3/7}$ in radical form.

   a) $\sqrt[3]{x^7}$ b) $\sqrt{x^{3/7}}$ c) $\sqrt[7]{x^3}$ d) $\sqrt[3]{x^{3/7}}$

Simplify. Assume all variables represent positive numbers.

3. $\sqrt[3]{64}$

   a) 8 b) 2 c) 16 d) 4

4. $\sqrt{96}$

   a) $4\sqrt{6}$ b) $6\sqrt{4}$ c) $16\sqrt{6}$ d) $8\sqrt{6}$

5. $\sqrt{18x^4}$

   a) $3x^2\sqrt{2}$ b) $(9x)^{1/3}$ c) $9x^{2/3}$ d) $2x^2\sqrt{3}$

6. $\sqrt{150x^5y^8}$

   a) $5x^2y^3\sqrt{6x}$ b) $5xy^4\sqrt{6x}$ c) $5x^2y^4\sqrt{6x}$ d) $5xy\sqrt{6xy}$

7. $\sqrt{16x^3y} \cdot \sqrt{48xy}$

   a) $16\sqrt{3x^4y^2}$ b) $16x^2y\sqrt{3}$ c) $16x^2y\sqrt{2}$ d) $16x^2y\sqrt{3x^2}$

8. $\sqrt{2a^2b^3} \sqrt{27a^5b}$

   a) $\sqrt{54a^7b^4}$ b) $3a^3b^2\sqrt{6a}$ c) $27a^3b^2\sqrt{6a}$ d) $9a^3b^2\sqrt{6a}$

9. $\sqrt{\dfrac{2}{32}}$

   a) $\dfrac{1}{\sqrt{4}}$ b) $\dfrac{1}{2}$ c) $\dfrac{\sqrt{2}}{\sqrt{32}}$ d) $\dfrac{1}{4}$

Practice Test 9B

10. $\dfrac{\sqrt{17x^3y}}{\sqrt{17y^5}}$

   a) $\dfrac{x^2\sqrt{x}}{y^2}$ b) $\dfrac{\sqrt{x}}{y^2}$ c) $\dfrac{x\sqrt{x}}{y^2}$ d) $\dfrac{17x\sqrt{x}}{y^2}$

11. $\dfrac{3}{\sqrt{6}}$

   a) 3 b) $\dfrac{\sqrt{18}}{6}$ c) $\dfrac{\sqrt{6}}{2}$ d) $\dfrac{3\sqrt{2}}{2}$

12. $\sqrt{\dfrac{81x}{5}}$

   a) $\dfrac{9\sqrt{5x}}{\sqrt{5}}$ b) $9\sqrt{x}$ c) $\dfrac{9\sqrt{5x}}{5}$ d) $\dfrac{\sqrt{45x}}{5}$

13. $\sqrt{\dfrac{20a^2b^3}{3a^3b^5}}$

   a) $\dfrac{2\sqrt{15a}}{3ab}$ b) $\dfrac{2\sqrt{15a}}{b\sqrt{3a}}$ c) $\dfrac{2}{b\sqrt{3a}}$ d) $\dfrac{2\sqrt{5}}{b}$

14. $\dfrac{3}{\sqrt{5}+\sqrt{6}}$

   a) $3\sqrt{5}-3\sqrt{6}$ b) $-3\sqrt{5}+3\sqrt{6}$ c) $-3\sqrt{5}-3\sqrt{6}$ d) $3\sqrt{5}+3\sqrt{6}$

15. $\dfrac{6}{\sqrt{x}-1}$

   a) $\dfrac{6\sqrt{x}-6}{x+1}$ b) $\dfrac{6\sqrt{x}+6}{x+1}$ c) $\dfrac{6\sqrt{x}+6}{x-1}$ d) $\dfrac{-6\sqrt{x}-6}{x+1}$

16. $5\sqrt{2}-3\sqrt{6}+\sqrt{6}-2\sqrt{2}+1$

   a) $3\sqrt{2}-2\sqrt{6}+1$ b) $\sqrt{8}+1$ c) $\sqrt{12}+1$ d) $2\sqrt{6}+\sqrt{2}+1$

17. $2\sqrt{a}-5\sqrt{a}+\sqrt{a}$

   a) $2\sqrt{a}$ b) $2a$ c) $-\sqrt{2a}$ d) $-2\sqrt{a}$

Practice Test 9B

Solve.

18. $\sqrt{2x+1} = 5$

   a) 0   b) 4   c) 2   d) 12

19. $5\sqrt{x} - 4 = x$

   a) 1, 25   b) 25, 16   c) 1, 16   d) 0, 4

20. Find the length indicated by x.

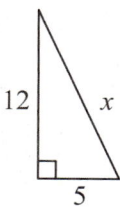

   a) 14   b) $\sqrt{119} \approx 10.91$ c) 13   d) $\sqrt{17} \approx 4.12$

21. Find the length of the line segment between the points (0, 3) and (−2, −5).

   a) $2\sqrt{17} \approx 8.25$ b) $2\sqrt{2} \approx 2.83$ c) $5\sqrt{2} \approx 7.07$ d) $\sqrt{29} \approx 5.39$

22. Evaluate $25^{-3/2}$.

   a) $\dfrac{1}{5}$ b) $\dfrac{1}{125}$ c) $\dfrac{1}{25}$ d) $\dfrac{1}{75}$

23. Simplify $\sqrt[4]{x^7} \cdot \sqrt[4]{x^5}$.

   a) $x^{48/35}$ b) x^3 c) x^2 d) $x^{35/16}$

24. An object is dropped from 100 feet. At what velocity does the object hit the ground? Use the formula $v = \sqrt{2gh}$ with $g = 32$ feet per second squared. The velocity will be in feet per second.

   a) 80 feet per second   b) 100 feet per second   c) 70 feet per second   d) 90 feet per second

25. If a square room has an area of 196 square feet, find the length of one side of the room.

   a) 13 feet   b) 15 feet   c) 14 feet   d) 12 feet

Notes:

Name:
Instructor:
Date:
Section:

10.1 The Square Root Property

Objectives
1. Know that every positive real number has two square roots.
2. Solve quadratic equations using the square root property.

Key Vocabulary
quadratic equation in standard form, square root property

1 Know that every positive real number has two square roots.

Example 1 Name the square roots of 25.

2 Solve quadratic equations using the square root property.

Example 2 Solve.

a) $x^2 = 169$

b) $x^2 - 7 = 74$

c) $a^2 - 20 = 0$

d) $(x+5)^2 = 64$

e) $(2x-5)^2 - 2 = 43$

Example 3 Use a quadratic equation to solve the problem.

Kathy decided to plant a rectangular garden so that it has the length-to-width ratio found in a golden rectangle. Determine the length and width of the garden if it is to have an area of 1800 square feet. (A rectangle whose length is about 1.62 times its width is called a golden rectangle.)

Answers: 1. ±5   2a) ±13   b) ±9   c) ±2√5   d) −13, 3   e) $\frac{5+3\sqrt{5}}{2}, \frac{5-3\sqrt{5}}{2}$ 3) width = $33\frac{1}{3}$ ft, length = 54 ft

Name:
Instructor:

Date:
Section:

Practice Set 10.1

Solve.

1. $x^2 - 225 = 0$

2. $a^2 - 18 = 82$

1._____

2._____

3. $5x^2 = 40$

4. $3y^2 + 7 = 28$

3._____

4._____

5. $(x-3)^2 = 81$

6. $(x+6)^2 = 45$

5._____

6._____

7. $(7x+6)^2 - 5 = 40$

8. $(3x+1)^2 + 2 = 50$

7._____

8._____

9. Kathy decided to plant a rectangular garden so that it has the length-to-width ratio found in a golden rectangle. Determine the length and width of the garden if it is to have an area of 1500 square feet. (A rectangle whose length is about 1.62 times its width is called a golden rectangle.)

9._____

_____

10. The product of two positive numbers is 180. Determine the numbers if the larger number is 5 times the smaller number.

10._____

Name:
Instructor:
Date:
Section:

10.2 Solving Quadratic Equations by Completing the Square

Objectives
1. Write perfect square trinomials.
2. Solve quadratic equations by completing the square.

Key Vocabulary
perfect square trinomial, completing the square

1 Write perfect square trinomials.

Example 1 Determine the number that must be added to create a perfect square trinomial.

a) $x^2 + 10x + \underline{}$

b) $y^2 - 6y + \underline{}$

2 Solve quadratic equations by completing the square.

Example 2 Solve by completing the square.

a) $x^2 + 5x + 6 = 0$

b) $x^2 = 16x - 48$

c) $a^2 + 14a = 4$

d) $x^2 - 7x - 12 = 0$

e) $2x^2 + 5x - 1 = 0$

f) $33x^2 + 6x = 6$

Example 3 Use a quadratic equation to solve the problem.

The product of a number and 1 less than half of the number is 40. Find the two positive numbers that satisfy this condition.

Answers: 1a) 25 b) 9   2a) −3, −2 b) 4, 12 c) $-7 \pm \sqrt{53}$ d) $\dfrac{7 \pm \sqrt{97}}{2}$ e) $\dfrac{-5 \pm \sqrt{33}}{4}$ f) $\dfrac{-1 \pm \sqrt{23}}{11}$ 3) 10, 4

Name:
Instructor:

Date:
Section:

Practice Set 10.2

Solve by completing the square.

1. $y^2 + 7y - 8 = 0$

2. $a^2 - 4a + 3 = 0$

1._____

2._____

3. $x^2 = 10x - 24$

4. $y^2 + 2y = 4$

3._____

4._____

5. $x^2 + 7x - 7 = 0$

6. $2a^2 + 9a - 1 = 0$

5._____

6._____

7. $28x^2 + 8x = 8$

8. $4y^2 - 20y - 20 = 0$

7._____

8._____

9. The product of a number and 1 less than half of the number is 40. Find the two negative numbers that satisfy this condition.

9._____

10. When an object is thrown straight up from a planet with an initial velocity of 96 feet per second, its height above the ground, s, in feet, after t seconds is given by the formula $s = -16t^2 + 96t$. How long after it is thrown will the height of the object be 80 feet? (Therefore, s = 80.)

10._____

Name:
Instructor:
Date:
Section:

10.3 Solving Quadratic Equations by the Quadratic Formula

Objectives
1. Solve quadratic equations by the quadratic formula.
2. Determine the number of solutions to a quadratic equation using the discriminant.

Key Vocabulary
quadratic formula, discriminant

1 Solve quadratic equations by the quadratic formula.

Example 1 Use the quadratic formula to solve each equation. If the equation has no real solution, so state.

a) $x^2 - 2x - 8 = 0$

b) $15y^2 = 2 - 7y$

c) $5x^2 - 7x - 3 = 0$

d) $5x^2 + 2x + 11 = 0$

e) $x^2 - 11x = -3$

Example 2 Use a quadratic equation to solve the problem.

The length of a rectangle is 4 inches less than twice the width. Find the length and width of the rectangle if the area is 336 square inches.

2 Determine the number of solutions to a quadratic equation using the discriminant.

Example 3 Determine whether each equation has two distinct real number solutions, one real number solution, or no real number solution.

a) $x^2 + 7x - 5 = 0$

b) $4y = 6 + y^2$

c) $a^2 - 6a + 9 = 0$

Answers: 1a) $-2, 4$ b) $\frac{1}{5}, -\frac{2}{3}$ c) $\frac{7 \pm \sqrt{109}}{10}$ d) no real number solution e) $\frac{11 \pm \sqrt{109}}{2}$ 2) length = 24 in., width = 14 in.
3a) two distinct real number solutions b) no real number solution c) one real number solution

Name:
Instructor:

Date:
Section:

Practice Set 10.3

Determine whether each equation has two distinct real number solutions, one real number solution, or no real number solution.

1. $x^2 + 5x - 3 = 0$
2. $4a = 12 + a^2$

1._____

2._____

3. $y^2 - 4y + 4 = 0$
4. $4x^2 + 20x + 25 = 0$

3._____

4._____

5. $(x-3)^2 = 81$
6. $(x+6)^2 = 45$

5._____

6._____

Use the quadratic formula to solve each equation. If the equation has no real solution, so state.

7. $x^2 - 3x - 40 = 0$
8. $y^2 - 4y - 5 = 0$

7._____

8._____

9. $3x^2 - 7x = 5$
10. $21a^2 = 2 - 11a$

9._____

10._____

11. $6x^2 + 3x + 12 = 0$
12. $y^2 - 9y + 1 = 0$

11._____

12._____

Practice Set 10.3

13. $x^2 + 3x - 5 = 0$

14. $3x^2 - 5x = 7$

13. _____

14. _____

15. When an object is thrown straight up from a planet with an initial velocity of 128 feet per second, its height above the ground, s, in feet after t seconds is given by the formula $s = -16t^2 + 128t$. How long after it is thrown will the height of the object be 213 feet? (Therefore, $s = 213$.)

15. _____

16. The diagonal of a slab of cement is 26 feet, and the length is 4 feet more than twice the width. Find the length and width of the slab.

16. _____
_____

Name:
Instructor:
Date:
Section:

10.4 Graphing Quadratic Equations

Objectives
1. Graph quadratic equations in two variables.
2. Find the coordinates of the vertex of a parabola.
3. Use symmetry to graph quadratic equations.
4. Find the intercepts of the graph of a quadratic equation.

Key Vocabulary
parabola, vertex, symmetry, axis of symmetry, x-intercept, y-intercept

1 Graph quadratic equations in two variables.

Example 1 Graph.

a) $y = x^2$

b) $y = x^2 + 1$

c) $y = -x^2 - 2x - 1$

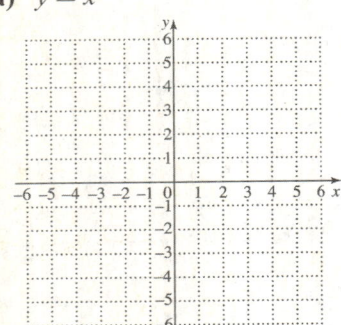 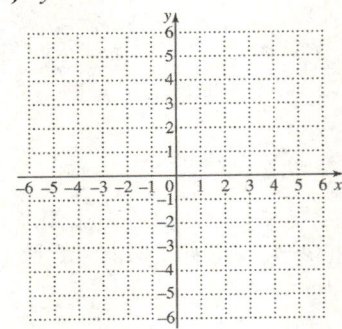 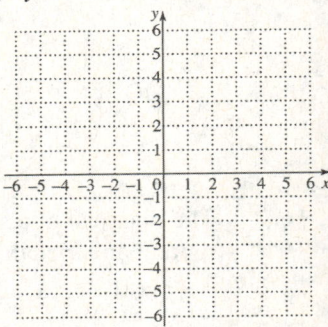

2 Find the coordinates of the vertex of a parabola.

Example 2 Indicate the coordinates of the vertex, and whether the parabola opens upward or downward.

a) $y = x^2 + 2x - 1$

b) $y = -2x^2 + 1$

c) $y = x^2 - 4x + 7$

Answers: 1a) 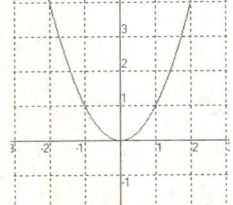 b) 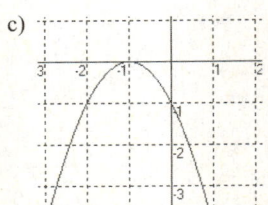 c)

2a) $(-1, -2)$, upward
b) $(0, 1)$, downward
c) $(2, 3)$, upward

Examples 10.4

3 Use symmetry to graph quadratic equations.

Example 3 Use the axis of symmetry to graph the equation.

a) $y = x^2 - 6x + 9$

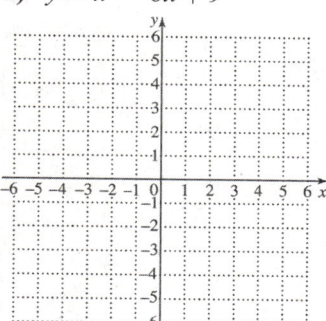

b) $y = x^2 - 6x + 10$

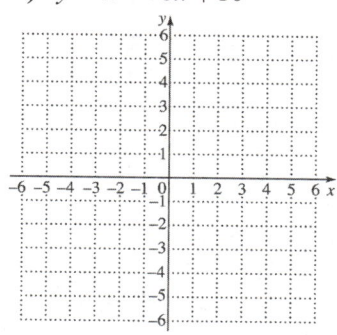

c) $y = -x^2 - 2x + 1$

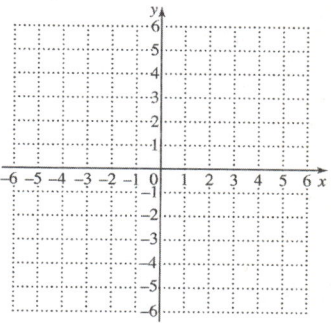

4 Find the intercepts of the graph of a quadratic equation.

Example 4 Use the intercepts to graph the equation.

a) $y = x^2 + 5x + 6$

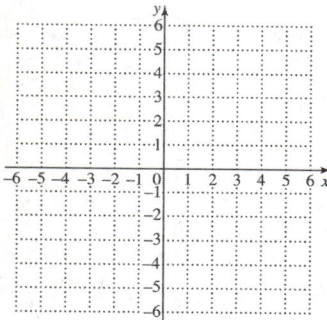

b) $y = -x^2 + x + 6$

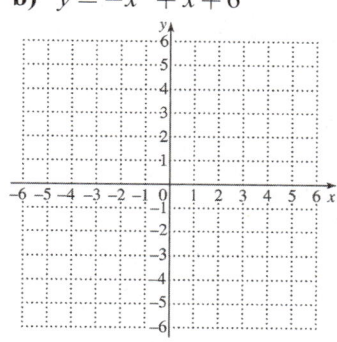

c) $y = x^2 + x - 2$

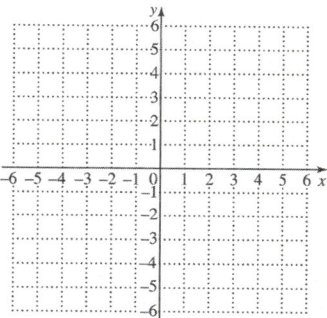

Answers: 3a) 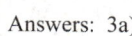 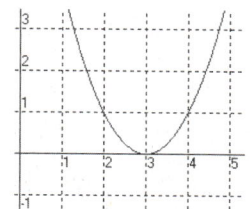 b) 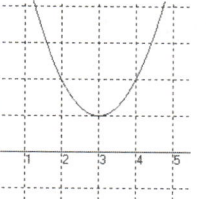 c) 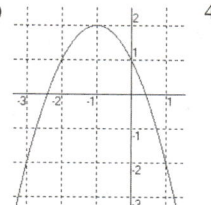 4a)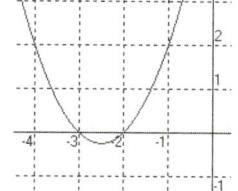

b) 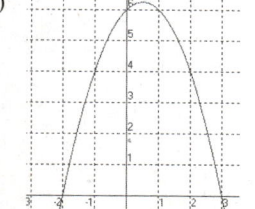 c)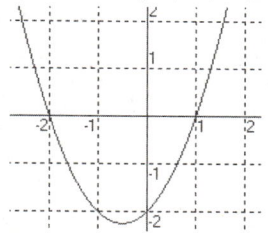

Name:
Instructor:

Date:
Section:

Practice Set 10.4

Indicate the axis of symmetry, the coordinates of the vertex, and whether the parabola opens upward or downward.

1. $y = -x^2 + 4$

2. $y = x^2 - 4x + 7$

3. $y = -2x^2 - 4x - 2$

4. $y = 3x^2 + 4x - 1$

1._____

2._____

3._____

4._____

Graph each quadratic equation and determine the x-intercepts, if they exist.

5. $y = x^2 - 3$

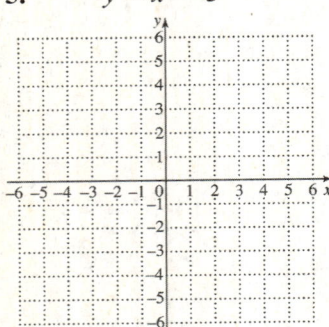

6. $y = -x^2 + 2$

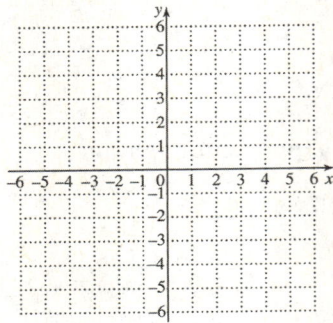

7. $y = x^2 - 4x + 6$

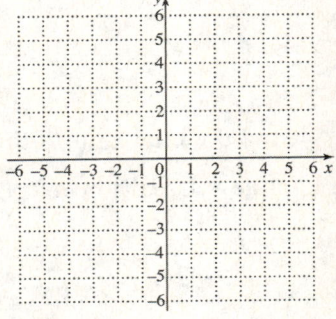

8. $y = x^2 + 6x + 8$

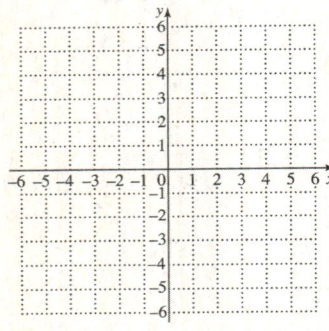

9. $y = 2x^2 - 12x + 17$

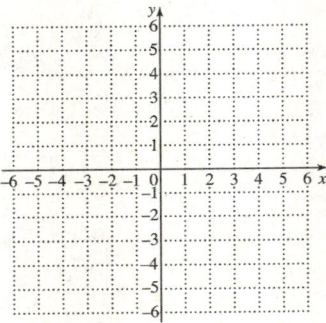

10. $y = -3x^2 - 6x - 4$

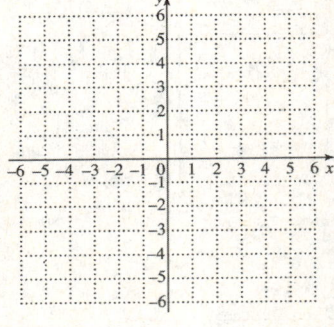

11. $y = x^2 - 4x$

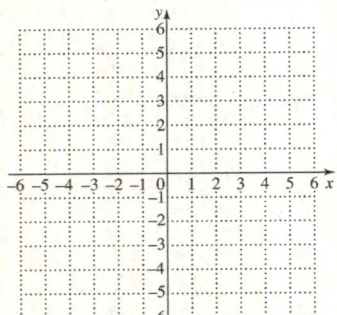

12. $y = -x^2 + 4x$

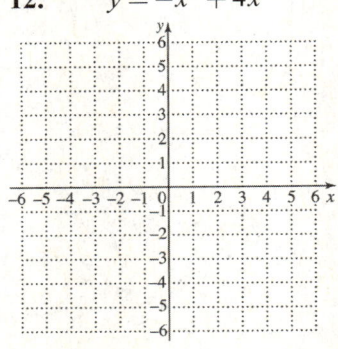

13. $y = x^2 + 2x + 5$

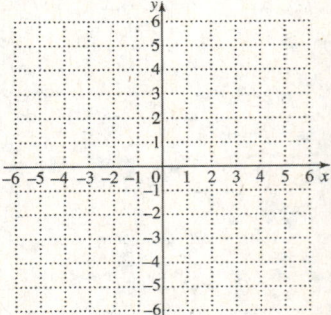

Name:
Instructor:

Date:
Section:

10.5 Complex Numbers

> **Objectives**
> 1. Write complex numbers using *i*.
> 2. Add and subtract complex numbers.
> 3. Solve quadratic equations with complex number solutions.
>
> **Key Vocabulary**
> imaginary numbers, imaginary unit, complex numbers, pure imaginary number

1 Write complex numbers using *i*.

Example 1 Write each of the following complex numbers in the form $a + bi$.

a) $\sqrt{-25}$

b) $\sqrt{-100}$

c) $3 + \sqrt{-16}$

d) $7 - \sqrt{-20}$

e) $5 + \sqrt{-24}$

2 Add and subtract complex numbers.

Example 2 Add or subtract the following complex numbers.

a) $(4 - 3i) + i$

b) $(2 + 3i) + (5 - 6i)$

c) $(6 - 5i) - (2 - 9i)$

3 Solve quadratic equations with complex number solutions.

Example 3 Solve the quadratic equation and write the solution in terms of *i*.

a) $x^2 = -81$

b) $2x^2 + 8 = 0$

c) $x^2 - 2x + 28 = 0$

d) $5x^2 - 3x + 10 = 0$

Answers: 1a) $5i$ b) $10i$ c) $3 + 4i$ d) $7 - 2i\sqrt{5}$ e) $5 + 2i\sqrt{6}$ 2 a) $4 - 2i$ b) $7 - 3i$ c) $4 + 4i$ 3 a) $\pm 9i$ b) $\pm 2i$ c) $1 \pm 3i\sqrt{3}$ d) $\dfrac{3 \pm i\sqrt{191}}{10}$

Name:
Instructor:

Date:
Section:

Practice Set 10.5

Write each imaginary number in terms of *i*.

1. $\sqrt{-49}$ 2. $\sqrt{-17}$ 3. $\sqrt{-54}$

1. _____

2. _____

3. _____

Write each complex number in the form of *a* + *bi*.

4. $5+\sqrt{-49}$ 5. $-3-\sqrt{-5}$ 6. $3-\sqrt{-98}$

4. _____

5. _____

6. _____

Add or subtract the following complex numbers.

7. $(3+7i)+8$ 8. $(2-5i)-2i$ 9. $(5+2i)+(-1+3i)$

7. _____

8. _____

9. _____

10. $(-7+3i)-(2-3i)$ 11. $(11-4i)-(11+4i)$

10. _____

11. _____

Solve the quadratic equation and write the solutions in terms of *i*.

12. $x^2=-25$ 13. $2y^2+64=0$

12. _____

13. _____

14. $x^2+2x+5=0$ 15. $a^2=4a-13$

14. _____

15. _____

16. $2x^2-6x+17=0$ 17. $12y^2-20y+11=0$

16. _____

17. _____

Chapter 10 Vocabulary Reference Sheet

Term	Definition	Example
quadratic equation in standard form	Equation of the form $ax^2 + bx + c = 0$ where a, b, and c are real numbers and $a \neq 0$	$2x^2 - 3x + \dfrac{1}{5} = 0$
square root property	If $x^2 = a$, then $x = \sqrt{a}$ or $x = -\sqrt{a}$ (abbreviated $x = \pm\sqrt{a}$).	If $x^2 = 3$, then $x = \sqrt{3}$ or $x = -\sqrt{3}$ (or $x = \pm\sqrt{3}$).
perfect square trinomial	A trinomial that can be expressed as the square of a binomial	$x^2 + 10x + 25 = (x+5)^2$
completing the square	A procedure for solving a quadratic equation in which a perfect square trinomial is produced on the left-hand side of the equation	$x^2 + 2x = 1$ $x^2 + 2x + 1 = 1 + 1$ $(x+1)^2 = 2$ $x + 1 = \pm\sqrt{2}$ $x = -1 \pm \sqrt{2}$
quadratic formula	$x = \dfrac{-b \pm \sqrt{b^2 - 4ac}}{2a}$	If $x^2 + 2x - 1 = 0$ then $x = \dfrac{-2 \pm \sqrt{2^2 - 4(1)(-1)}}{2 \cdot 1}$ $x = \dfrac{-2 \pm \sqrt{4+4}}{2}$ $x = \dfrac{-2 \pm 2\sqrt{2}}{2}$ $x = \dfrac{2(-1 \pm \sqrt{2})}{2} = -1 \pm \sqrt{2}$
discriminant	The expression under the square root sign $(b^2 - 4ac)$ in the quadratic formula, which can be used to determine the number of real solutions to a quadratic equation	$x^2 - 10x + 24 = 0$: The discriminant is $(-10)^2 - 4(1)(24) = 4$. Therefore, the equation has two real number solutions. $x^2 + 6x + 9 = 0$: The discriminant is $(6)^2 - 4(1)(9) = 0$. Therefore, the equation has one real number solution. $3x^2 + 5x + 8 = 0$: The discriminant is $(5)^2 - 4(3)(8) = -71$. Therefore, the equation has no real number solution.
parabola	The graph of every quadratic equation	$y = ax^2 + bx + c$, $a > 0$ $y = ax^2 + bx + c$, $a < 0$

Chapter 10 Vocabulary

vertex of a parabola	The lowest point on a parabola that opens upward or the highest point on a parabola that opens downward. The coordinates of the vertex are $\left(-\dfrac{b}{2a},\, \dfrac{4ac-b^2}{4a}\right).$	The vertex of the parabola $y = x^2 - 6x + 4$ is $\left(-\dfrac{-6}{2\cdot 1},\, \dfrac{4(1)(4)-(-6)^2}{4\cdot 1}\right) = (3,\, -5).$
axis of symmetry	A vertical line through the vertex of the parabola where the left and right sides have symmetry	For $y = x^2 - 6x + 4$, the axis of symmetry is the line $x = -\dfrac{b}{2a}$ or $x = -\dfrac{-6}{2\cdot 1} = 3.$
x-intercept	A point where a graph crosses the x-axis; $(x, 0)$	
y-intercept	A point where a graph crosses the y-axis; $(0, y)$	
imaginary unit	$i = \sqrt{-1}$	
imaginary numbers	A number of the form $a + bi$, where $b \neq 0$ and i is the imaginary unit	$-5i$, $3+2i$
complex numbers	Every number of the form $a + bi$, where a and b are real numbers and i is the imaginary unit	$-4i$, $5 + 2i$, 2.5, $-\dfrac{1}{3}$, 0, $3 - i\sqrt{2}$, $1.\overline{3}$
pure imaginary number	A number of the form $a + bi$, where $a = 0$ and i is the imaginary unit	$9i$, $-i\sqrt{7}$

Name:
Instructor:

Date:
Section:

Chapter 10 Practice Test A

1. Solve $2x^2 - 3 = 37$ using the square root property.

    1._____

2. Solve $(4x+5)^2 = 10$ using the square root property.

    2._____

3. Solve $x^2 - 8x - 2 = 0$ by completing the square.

    3._____

4. Solve $2x^2 - 10x + 12 = 0$ by completing the square.

    4._____

5. Solve $8x^2 - 10x = -3$ by the quadratic formula.

    5._____

6. Solve $3x^2 + 7x - 4 = 0$ by the quadratic formula.

    6._____

7. Solve $(x-5)^2 = 75$ by the method of your choice.

    7._____

8. Give an example of a perfect square trinomial.

    8._____

9. Determine whether $-2x^2 + 3x + 1 = 0$ has two distinct real number solutions, one real number solution, or no real number solution.

    9._____

Practice Test 10A

10. Determine whether $-2x^2 + 5x - 4 = 0$ has two distinct real solutions, one real solution, or no real solution.

10._____

11. Find the equation of the axis of symmetry of the graph of $y = -x^2 + 2x + 3$.

11._____

12. Find the equation of the axis of symmetry of the graph of $y = x^2 - 6x + 10$.

12._____

13. Determine whether the graph of $y = 3x^2 - 2x + 1$ opens upward or downward.

13._____

14. Determine whether the graph of $y = -3x^2 + 5x$ opens upward or downward.

14._____

15. Find the vertex of the graph of $y = x^2 - 2x + 5$.

15._____

16. Find the vertex of the graph of $y = -2x^2 + 8x - 6$.

16._____

17. Graph the equation $y = x^2 - 3x$ and determine the *x*-intercepts, if they exist.

17._____

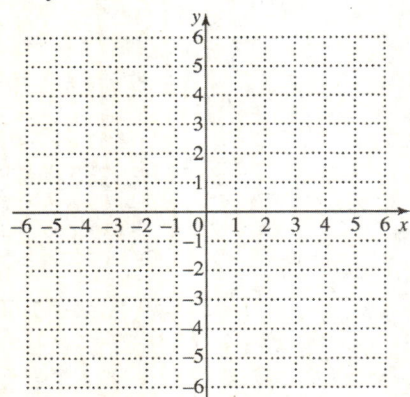

18. Graph the equation $y = -x^2 + 5x - 6$ and determine the x-intercepts, if they exist.

18._____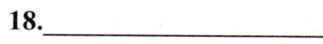

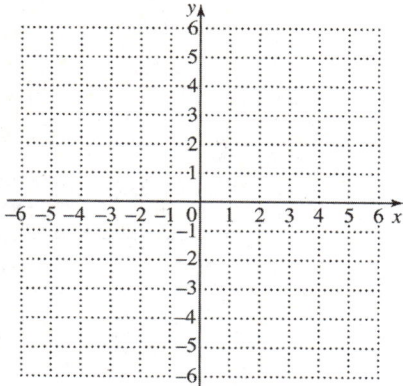

19. Graph the equation $y = x^2 - 4x + 1$ and determine the x-intercepts, if they exist.

19._____

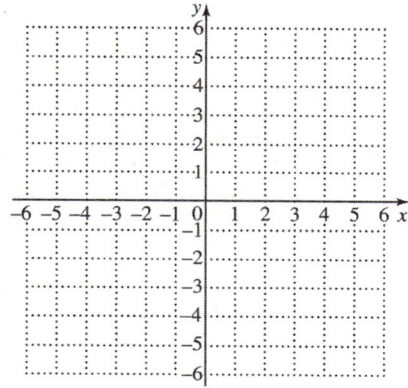

20. The product of two consecutive odd integers is 195. Find the two integers.

20._____

21. An object is thrown upward from the ground. The height, s, of the object above the ground at t seconds after it is thrown can be found by the formula $s = -16t^2 + 60t$. Find how long it will take the object to hit the ground.

21._____

Practice Test 10A

22. Simplify $(3-2i)-(5+i)+\sqrt{-4}$ and write the result in standard form.

22._____

23. Simplify $-3-\sqrt{-9}$.

23._____

24. Solve the quadratic equation $(x+5)^2 = -49$ and write the answer in terms of i.

24._____

25. Solve the quadratic equation $4x^2 - 20x + 79 = 0$ and write the answer in terms of i.

25._____

Chapter 10 Practice Test B

1. Solve $2x^2 - 3 = 105$ using the square root property.

   a) $\pm 3\sqrt{6}$ b) $\pm 9\sqrt{6}$ c) $\pm 6\sqrt{3}$ d) $\pm 4\sqrt{3}$

2. Solve $(2x+5)^2 = 10$ using the square root property.

   a) $-5 \pm \sqrt{5}$ b) $\dfrac{-5 \pm \sqrt{10}}{2}$ c) $\dfrac{5 \pm \sqrt{10}}{2}$ d) $5 \pm \sqrt{5}$

3. Solve $x^2 - 2x - 1 = 0$ by completing the square.

   a) $-1 \pm \sqrt{2}$ b) $1 \pm \sqrt{2}$ c) $\pm \sqrt{2}$ d) $2 \pm \sqrt{2}$

4. Solve $3x^2 - 12x + 7 = 0$ by completing the square.

   a) $2 \pm \sqrt{15}$ b) $\dfrac{6 \pm \sqrt{5}}{3}$ c) $2 \pm \sqrt{5}$ d) $\dfrac{6 \pm \sqrt{15}}{3}$

5. Solve $2x^2 + 2x = 24$ by the quadratic formula.

   a) $-2, 1$ b) $-4 \pm \sqrt{3}$ c) $-4, 3$ d) $-\dfrac{1}{4}, \dfrac{1}{3}$

6. Solve $2x^2 + 6x + 1 = 0$ by the quadratic formula.

   a) $-3 \pm \dfrac{\sqrt{7}}{2}$ b) $-\dfrac{3}{2} \pm 2\sqrt{7}$ c) $\dfrac{-3 \pm \sqrt{7}}{2}$ d) $-\dfrac{5}{2} \pm \sqrt{7}$

7. Solve $(x-5)^2 = 20$ by the method of your choice.

   a) $5 \pm 2\sqrt{5}$ b) $\pm 7\sqrt{5}$ c) $\pm 3\sqrt{5}$ d) $2 \pm \sqrt{5}$

8. Which of the following is an example of a perfect square trinomial?

   a) $x^2 + 2x - 1$ b) $x^2 + 3x - 10$ c) $x^2 + 6x + 9$ d) $x^2 + 5x + 6$

9. Determine whether $x^2 - 4x + 4 = 0$ has two distinct real number solutions, one real number solution, or no real number solution.

   a) two distinct real number solutions   b) one real number solution   c) no real number solution

10. Determine whether $4x^2 - 3x - 2 = 0$ has two distinct real number solutions, one real number solution, or no real number solution.

    a) two distinct real number solutions   b) one real number solution   c) no real number solution

Practice Test 10B

11. Find the equation of the axis of symmetry of the graph of $y = -x^2 - 6x - 10$.

a) $x = -3$ b) $x = 3$ c) $x = 1$ d) $x = -1$

12. Find the equation of the axis of symmetry of the graph of $y = x^2 - 2x + 4$.

a) $x = -3$ b) $x = 3$ c) $x = 1$ d) $x = -1$

13. Determine whether the graph of $y = x^2 - x - 10$ opens upward or downward.

a) upward b) downward

14. Determine whether the graph of $y = -2x^2 + 5x + 1$ opens upward or downward.

a) upward b) downward

15. Find the vertex of the graph of $y = x^2 - 4x + 8$.

a) (2, 4) b) (−2, 4) c) (2, −4) d) (−2, −4)

16. Find the vertex of the graph of $y = -x^2 - 8x - 9$.

a) (4, 7) b) (−4, 7) c) (4, −7) d) (−4, −7)

17. Graph the equation $y = x^2 - 2x$.

a) 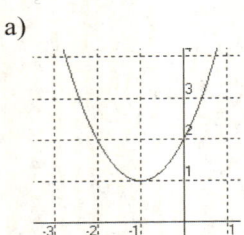 b) 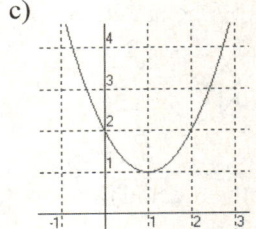 c) 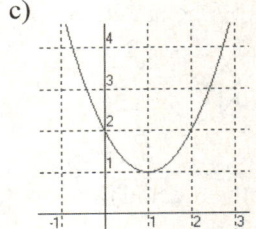 d)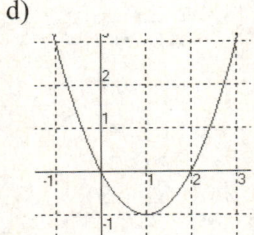

18. Graph the equation $y = -x^2 + 2x + 1$.

a) 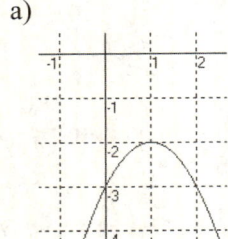 b) 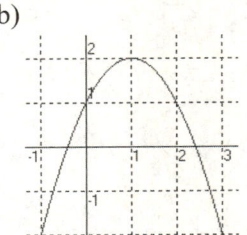 c) 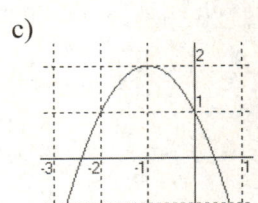 d)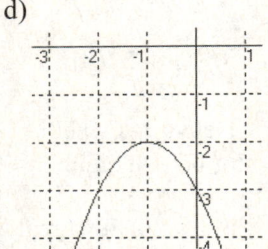

Practice Test 10B

19. Graph the equation $y = x^2 - 4x + 5$.

a) 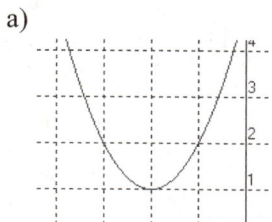   b) 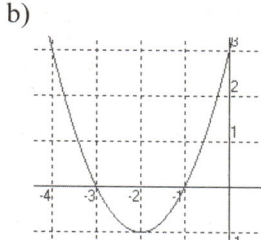   c) 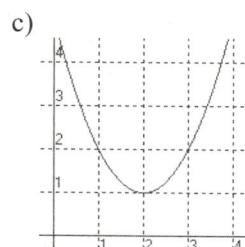   d)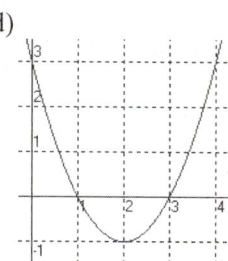

20. The product of two consecutive even integers is 168. Find the two integers.

   a) 8, 21   b) 10, 12   c) 12, 14   d) 14, 16

21. An object is thrown upward from the ground. The height, s, of the object above the ground at t seconds after it is thrown can be found by the formula $s = -16t^2 + 64t$. Find how long it will take the object to hit the ground.

   a) 4 seconds   b) 3 seconds   c) 3.5 seconds   d) 4.5 seconds

22. Simplify $(3 - 2i) - (7 + i) + \sqrt{-9}$ and write the result in standard form.

   a) $-4 + i$ b) $-4 - i$ c) -4 d) $-4 + 2i$

23. Simplify $-5 - \sqrt{-25}$.

   a) 0   b) $-5 - 5i$ c) $-10i$ d) i

24. Solve the quadratic equation $(x + 3)^2 = -25$ and write the answer in terms of i.

   a) $\pm 8i$ b) $\pm 2i$ c) $3 \pm 5i$ d) $-3 \pm 5i$

25. Solve the quadratic equation $2x^2 + x + 3 = 0$ and write the answer in terms of i.

   a) $\dfrac{-1 \pm i\sqrt{23}}{4}$ b) $\dfrac{1 \pm i\sqrt{23}}{2}$ c) $-\dfrac{1}{4} \pm i\sqrt{23}$ d) $-1 \pm \dfrac{i\sqrt{23}}{4}$

Notes:

Chapter 1 Answers

Section 1.2
1. 3, 2, 1, 5, 4
3. a) soybeans; 34 million acres
   b) canola; 2 million acres
5. a) $87.49   b) $85.13
7. $837.90
9. a) $\frac{8}{14}$ or $\frac{4}{7}$ b) 57%   c) 14%
11. a) Sun.   b) Mon. & Fri.
    c) 16 h

Section 1.3
1. Variables
3. fraction
5. LCD
7. 4
9. $\frac{5}{6}$
11. $\frac{1}{3}$
13. $\frac{83}{72}$ or $1\frac{11}{72}$
15. $\frac{1}{4}$
17. $\frac{7}{5}$ or $1\frac{2}{5}$
19. $\frac{58}{7}$
21. a) 2 cups
    b) $4\frac{1}{2}$ tsp paprika, 1 cup parsley, 2 cups cilantro, $1\frac{1}{2}$ tsp cumin, $\frac{3}{8}$ tsp cayenne

Section 1.4
1. E
3. E
5. {32}
7. {0, 32}
9. {$\sqrt{10}$}
11. F
13. F
15. T
17. Answers may vary.
19. not possible

Section 1.5
1. <
3. >
5. >
7. <
9. >
11. 6
13. −13
15. >
17. <
19. <

Section 1.6
1. positive
3. additive inverse, opposite
5. additive inverse, opposite
7. a) positive   b) 0.08
9. a) negative   b) $-\frac{1}{6}$
11. −15
13. −0.6
15. −18
17. −68
19. $-\frac{7}{4}$
21. $\frac{1}{16}$

Section 1.7
1. −1
3. 0
5. −42
7. 42
9. −13
11. $\frac{7}{60}$
13. $-\frac{11}{16}$
15. $-\frac{19}{21}$
17. $-\frac{11}{35}$
19. 0

Section 1.8
1. zero
3. positive
5. positive
7. −27
9. −48
11. 1
13. −112
15. $-\frac{4}{5}$
17. $-\frac{49}{81}$
19. 10
21. 6
23. $\frac{4}{9}$
25. −6.35
27. 0
29. −9°F

Section 1.9
1. 64
3. 64
5. a) positive   b) 9
7. a) positive   b) 1
9. a) negative   b) −27
11. a) negative   b) −1
13. 21
15. $-\frac{25}{3}$
17. 6
19. −20
21. 81
23. $\frac{3}{4}$

Section 1.10
1. 1
3. 3
5. commutative property of addition
7. commutative property of multiplication
9. multiplicative inverse
11. commutative property of addition
13. commutative property of multiplication
15. distributive
17. distributive
19. distributive

Chapter 1 Answers

Practice Test 1A
1. $808.11
2. 4.9% discount; she will pay a total of $14,500 with the $500 rebate, but only 14,265 with the 4.9% discount.
3. 12,210 cars/day
4. a) 95 b) 29 c) 18
5. $\left\{-\dfrac{1}{2}, \dfrac{2}{3}, -5, 0, 98, 2.5\right\}$
6. $28.80
7. <
8. <
9. $-\dfrac{1}{9}$
10. $\dfrac{9}{10}$
11. $\dfrac{11}{60}$
12. $-\dfrac{37}{120}$
13. −8
14. −12
15. $\dfrac{1}{81}$
16. −4
17. 61
18. −6
19. distributive
20. associative property of addition

Practice Test 1B
1. d
2. d
3. a
4. b
5. b
6. a
7. c
8. b
9. b
10. d
11. c
12. b
13. d
14. b
15. b
16. d
17. d
18. b
19. d
20. a

Chapter 2 Answers

Section 2.1
1. expression
3. coefficient
5. combine
7. $-7x-5$
9. $1.3x+9.1$
11. $-x^2-3xy+2x$
13. $2x^2+4x$
15. $-\dfrac{2}{3}+2x$
17. $200x-4$
19. $13x-35$
21. -27

Section 2.2
1. D
3. E
5. B
7. 10
9. 0
11. 0
13. 3
15. -20
17. 0
19. 12

Section 2.3
1. 7
3. $\dfrac{3}{8}$
5. -15
7. $\dfrac{1}{21}$
9. -160
11. $-\dfrac{1}{7}$
13. -75
15. $-\dfrac{3}{11}$
17. 0
19. -6

Section 2.4
1. D or E
3. A
5. B
7. 2
9. -3
11. -7
13. $\dfrac{1}{3}$
15. -4
17. $0.08\overline{1}$

Section 2.5
1. C
3. A
5. 3
7. 4
9. 0.1
11. $\dfrac{6}{11}$
13. 3
15. $-\dfrac{35}{6}$
17. $\dfrac{1}{2}$
19. -6

Section 2.6
1. trapezoid, quadrilaterals
3. perimeter, circumference
5. 21.6
7. 78.4
9. $y=\dfrac{2}{3}x-4$
11. $h=\dfrac{3V}{\pi r^2}$
13. $52.50
15. 11 ft^2
17. 12.25 ft^2
19. a) $297.50
    b) $2297.50

Section 2.7
1. C
3. E
5. $2:5$
7. $1:2$
9. $0.4:1$
11. $\dfrac{5}{8}$
13. 175 mi
15. 162.56 cm
17. $\dfrac{5}{8}$ cup

Section 2.8
1. C
3. B
5. A
7. $x\geq 8$, $[8,\infty)$

9. All real numbers, $(-\infty,\infty)$

11. $x>\dfrac{4}{5}$, $\left(\dfrac{4}{5},\infty\right)$

13. $m\leq -7$, $(-\infty,-7]$

15. $n\leq 30$, $(-\infty,30]$

17. $x>5$, $(5,\infty)$

Chapter 2 Answers

Practice Test 2A
1. $-15x + 12$
2. $-13x - y + 4$
3. $-\dfrac{1}{10}x + \dfrac{7}{24}$
4. $-4x^2 + 17x - 6$
5. -2
6. all real numbers
7. -0.1125
8. 140
9. $y = 6x - 3$
10. $h = \dfrac{2A}{(B+b)}$
11. $x > -\dfrac{13}{2}$, $\left(-\dfrac{13}{2}, \infty\right)$

    ←−+−⊕−+−+→
       −13/2

12. $x \geq -\dfrac{35}{18}$, $\left[-\dfrac{35}{18}, \infty\right)$

    ←−+−●−+−+→
       −35/18

13. 2
14. 1 tsp
15. 800 in.³, yes
16. 38 pecans
17. 48 min
18. $569.25

Practice Test 2B
1. b
2. a
3. d
4. c
5. a
6. d
7. d
8. a
9. b
10. d
11. c
12. d
13. a
14. b
15. c
16. b
17. c
18. b

Chapter 3 Answers

Section 3.1
1. B
3. F
5. $2p$
7. $\frac{1}{2}w$
9. $\frac{2}{3}c - 9$
11. a) x = average test score
    b) $x - 12$
13. a) x = cost of Kyle's dinner
    b) $54.35 - x$
15. $(8 + 10x) + x = 41$
17. $(80 + 2x) + x = 845$

Section 3.2
1. D
3. F
5. 20
7. 178 and 180
9. 10 pages
11. $1444.44
13. 420 books

Section 3.3
1. D
3. E
5. C
7. $29°, 61°$
9. $11°, 11°$
11. 27 ft by 27 ft
13. $50°, 60°, 70°$

Section 3.4
1. $0.5x - 1$; $2x$
3. Cody 4 h; Jacob 5 h
5. $2\frac{1}{4}$ mi
7. $3\frac{1}{2}$ mi
9. $1\frac{1}{2}$ lb of cashews
   $4\frac{1}{2}$ lb of raisins
11. a) $4000 at $3\frac{1}{2}$%;
       $2000 at 4%
    b) $2666.67 at $3\frac{1}{2}$%;
       $3333.33 at 4%
    c) $3200 at 3 1/2 %;
       $2800 at 4 %

Practice Test 3A
1. $4700 - m$
2. $24d$
3. $s + 0.06s$ or $1.06s$
4. $x =$ number of pens; $x - 7$
5. $x =$ cost of store-brand macaroni and cheese; $0.49 + x$
6. 122, 123
7. 43, 83
8. $46.80
9. 37.5 in., 18.75 in., 19.75 in.
10. $50°, 100°, 30°$
11. 12 in. by 20 in.
12. 4.4 min
13. 36 mi
14. 4 main-floor tickets, 8 balcony tickets
15. 80 gal

Practice Test 3B
1. d
2. b
3. a
4. c
5. d
6. c
7. b
8. c
9. a
10. c
11. b
12. d
13. a
14. a
15. c

Chapter 4 Answers

Section 4.1
1. coefficient
3. add
5. denominator, numerator
7. factor
9. $(2x)^5$
11. w^{12}
13. $6b^4$
15. $\dfrac{81v^2}{t^6}$
17. $-9800s^{12}$
19. $-72b^2c^{13}d^8$
21. $27x^6y^{12}$ cm³

Section 4.2
1. false, $\dfrac{1}{3\cdot 3\cdot 3}=\dfrac{1}{27}$
3. true
5. true
7. false, 1
9. true
11. $\dfrac{n^2}{m^3}$
13. $\dfrac{49}{9}$
15. $\dfrac{1}{225}$
17. -2
19. -3

Section 4.3
1. E
3. C
5. B
7. 23,900
9. 7.5
11. 2.52×10^7
13. 1.35×10^6
15. 5.0×10^{-5}
17. 2.0×10^{-1}
19. 8.75×10^2 kWh

Section 4.4
1. 1, $-3m$, 1
3. 4, $p^3-7p^2+\dfrac{1}{2}p-9$, 3
5. 2, p^5+6, 5

7. $8x^2+7x+2$
9. y^2
11. $-2.4z^2+3.2z-9$
13. $8x^2-x+10$
15. $-a^2+2a+2$
17. $8p^2+4p+9$
19. $(10a-5)$ cm

Section 4.5
1. C
3. A
5. E
7. A
8. B
9. $-5m^4n^3$
11. $9t^5v-9t^2v^3$
13. $14d^2-55df-36f^2$
15. $4w^2-\dfrac{1}{9}$
17. $24b^3+11b^2-10b+24$
19. $-5s^2+28s+12$

Section 4.6
1. $3x-2$
3. $4x+5-\dfrac{2}{3x-2}$
5. $6c+2$
7. $3t^2-\dfrac{7}{3}t-2+\dfrac{5}{3t}$
9. $2x+1$
11. $p+3$
13. $2a^2+a+1+\dfrac{2}{3a-1}$
15. $8h-10$
17. $36s^2+42s+49$

Practice Test 4A
1. m^8
2. c^{18}
3. $-\dfrac{27t^6s^3}{r^9}$
4. $\dfrac{4a^4}{b^2}$
5. $\dfrac{1}{64}$
6. m^4

7. $30p^2r^5$
8. $\dfrac{u^3}{3v^4}$
9. 0.0000000132
10. 9,700,000,000
11. 0.13
12. 0.000006
13. 2.6×10^7
14. 1.26×10^2
15. $-n^2+3n+2$
16. $1.5w+0.1$
17. $-16y-11$
18. $6t^2-9t+5$
19. $6a^2-6$
20. $-22f^3-10f^2+18f$
21. $g-8+\dfrac{4}{g}$
22. $x-2$
23. 2160 mi
24. $(18x-3)$ in.

Practice Test 4B
1. c
2. b
3. c
4. c
5. b
6. d
7. a
8. b
9. b
10. b
11. b
12. a
13. b
14. c
15. d
16. a
17. b
18. b
19. c
20. c
21. d
22. b
23. a
24. b
25. d

Chapter 5 Answers

Section 5.1
1. $2^3 \cdot 3 \cdot 7$
3. $2^2 \cdot 5^3$
5. 1
7. $w-2$
9. $7(p-5)$
11. $9p(p^2-3p-4)$
13. $(3h-1)(7h+1)$
15. $r(5r-1)(4r-7)$

Section 5.2
1. $(p+5)(p+12)$
3. $(h+3)(h-15)$
5. $(f-10)(f-12)$
7. $9(t+2) - s(t+2) =$
   $(9-s)(t+2)$
9. $c(b-8) - 5(b-8) =$
   $(c-5)(b-8)$

Section 5.3
1. $(a-6b)(a+2b)$
3. $(t-10)(t+5)$
5. $(s+13)(s-2)$
7. $(k-6)(k+1)$
9. $3(c+7)(c+3)$
11. $-1(f-6)(f-9)$
13. $-2n(n+7)(n+10)$
15. $(r-17)(r-6)$
17. $7w^3(w+8)(w-7)$
19. $(y-20)(y+8)$

Section 5.4
1. $(5v+1)(4v-1)$
3. $(r-1)(8r-5)$
5. $(2k-5)(5k-2)$
7. $(t+7)(7t-2)$
9. $(11b+1)(b-5)$
11. $(2r-5)^2$
13. $(12y-z)(y-12z)$
15. $(2p+5)(11p-2)$

17. $(3d+4)(8d+3)$
19. $(3u-4)(6u-5)$

Section 5.5
1. $7s^2(5s-4)$
3. $(m+4n)(m^2-4mn+16n^2)$
5. prime
7. $(k+h)(2+j)$
9. $(11z-1)^2$
11. $(p^2+3)(p+2)$
13. $4u(2u+1)(u-5)$
15. $3mn(2m+5n)(2m-5n)$
17. $(x+y)(m+3a)$
19. $6(a+3b)(a-3b)$

Section 5.6
1. $\dfrac{3}{2}, 5$
3. $-4, 2$
5. $-\dfrac{1}{2}, 3$
7. $\dfrac{5}{6}, 1$
9. $-3, -\dfrac{1}{3}$
11. $\dfrac{1}{2}, -3$
13. $8, -6$
15. $1, -1$
17. $6, -4$
19. $7, 1$

Section 5.7
1. $-12, -7$
3. 4 units
5. 3 and 15
7. 25 ft
9. 3 in., 4 in., 5 in.

Practice Test 5A
1. $4a$
2. 1
3. $24y^2(3y-2)$
4. $13rs(1-7rs)$

5. $(t+u)(2t-3)$
6. $(6w-5)(w+7)$
7. $(h+18)(h+4)$
8. $(m-14)(m+4)$
9. prime
10. $(1+5b)(1-5b+25b^2)$
11. $(3c-7)(2c+5)$
12. $(5k+3)^2$
13. $(14n+1)(14n-1)$
14. prime
15. $4(y-2z)(y^2+2yz+4z^2)$
16. $\dfrac{3}{2}, -\dfrac{3}{2}$
17. $-\dfrac{1}{3}, \dfrac{3}{2}$
18. $9, -7$
19. $-4, 2$
20. James is 7; Samuel is 19.
21. 14, 16
22. 18 ft

Practice Test 5B
1. c
2. b
3. a
4. b
5. b
6. a
7. c
8. d
9. b
10. c
11. d
12. c
13. c
14. a
15. b
16. b
17. c
18. d
19. d
20. b

Chapter 6 Answers

Section 6.1
1. C
3. B
5. $\dfrac{3b^2}{2a}$
7. $\dfrac{1}{3c+1}$
9. $\dfrac{y-7}{x-1}$
11. $\dfrac{1}{b-2}$
13. $\dfrac{-1}{s+3}$
15. $\dfrac{(2x+5)(x-3)}{(x+3)(2x-1)}$
17. $\dfrac{2(x+5)}{x-5}$

Section 6.2
1. $\dfrac{3}{2y}$
3. $\dfrac{2(a-1)}{a-6}$
5. $\dfrac{b^2}{a}$
7. $\dfrac{(3x+5)(3x-2)}{(9x-1)(2x+1)}$
9. $\dfrac{25x}{4y^2}$
11. $\dfrac{(a-1)^3}{8(a-6)}$
13. $\dfrac{9x}{2(x+7)(x+2)}$
15. $\dfrac{x-1}{5x+1}$

Section 6.3
1. $\dfrac{5}{x}$
3. $\dfrac{3a}{b}$
5. $\dfrac{-rs}{r+2}$
7. $\dfrac{4}{x+1}$
9. $\dfrac{-1}{2(x+2)}$
11. C
13. B
15. D
17. 0

Section 6.4
1. $\dfrac{3a+b}{ab}$
3. $\dfrac{2-x^3}{5x^2}$
5. $\dfrac{3x+y}{x+y}$
7. $\dfrac{3x^2+5x+8}{(x-1)^2(x+1)}$
9. $\dfrac{-x-9}{9x}$
11. $\dfrac{10r+4}{(2r+1)(3r+2)(r-2)}$
13. $\dfrac{2x+7}{9(x+3)(x-3)}$
15. $\dfrac{-3(x+4)}{4x}$
17. $\dfrac{1}{2(\Delta^2-4)}$

Section 6.5
1. $\dfrac{b+3a}{3a(3b-1)}$
3. $\dfrac{7-3x}{14x}$
5. $\dfrac{y+16}{y+6}$
7. -1
9. $\dfrac{xy^2}{2y^2-3x^2}$
11. $-\dfrac{m}{n}$
13. $\dfrac{b+a}{b-a}$

Section 6.6
1. no solution
3. 2
5. no solution
7. 0
9. no solution
11. 12
13. $\dfrac{7}{3}$

Section 6.7
1. $\dfrac{5}{x-2}, \dfrac{7}{x}$
3. 4 mph
5. $w=2$ ft, $l=10$ ft
7. 8.8 mph
9. $2\dfrac{2}{9}$ hr
11. $13\dfrac{1}{3}$ hr or 13 hr 20 min

Section 6.8
1. A
3. B
5. A
7. 0.005
9. 10
11. $\dfrac{8}{9}$
13. 23 rugs
15. 1.4 in.

Chapter 6 Answers

Practice Test 6A

1. $\dfrac{2}{x+5}$
2. $-\dfrac{2}{y}$
3. $x \neq -1$
4. $x \neq 6, x \neq -6$
5. $\dfrac{9}{b^2 c}$
6. $x+1$
7. $\dfrac{2x+1}{x+1}$
8. $-\dfrac{1}{4}$
9. $\dfrac{5x+3}{6x}$
10. 2
11. $\dfrac{10-11xy^2}{6x^2 y}$
12. $\dfrac{x}{14(x-1)(x+1)}$
13. $\dfrac{x^2+9x+12}{(x+3)(x-2)(x+2)}$
14. $\dfrac{2(x+2)(x-1)}{(x+1)(x+3)}$
15. 8
16. $\dfrac{4y-x}{11x}$
17. -2
18. 1
19. -4
20. no solution
21. 72 min or 1 hr 12 min
22. 4 or $\dfrac{1}{4}$
23. 4 hr 10 min

Practice Test 6B

1. c
2. c
3. d
4. b
5. c
6. b
7. d
8. d
9. a
10. d
11. a
12. a
13. b
14. d
15. b
16. b

Chapter 7 Answers

Section 7.1
1. II
3. II
5. IV
7. a) $(1, 3)$
   b) $(-4, 1)$
   c) $(3, -2)$
   d) $(-2, -2)$
   e) $(0, -2)$
   f) $\left(\dfrac{3}{2}, 0\right)$
9.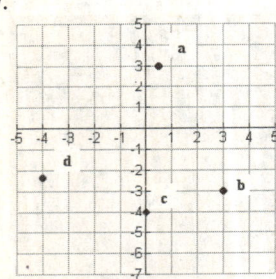
11. $(2, 3), (4, 0)$
13. $y = 1, \ x = 2$
15. a) $x < 0$
    b) $y > 0$

Section 7.2
1.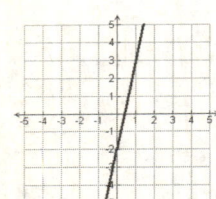
3.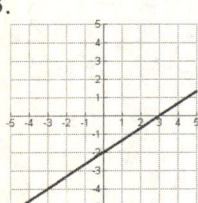
5.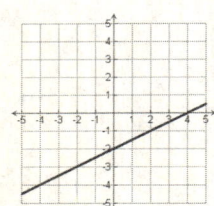
7.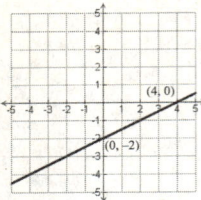
9.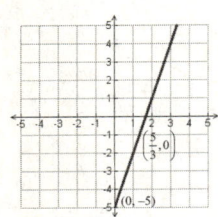
11.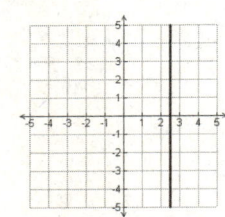
13. $y = 1$
15. a)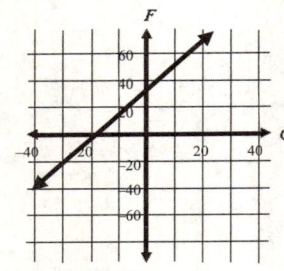
    b) $\approx 60°\text{F}$
    c) $\approx -7°\text{C}$

Section 7.3
1. A
3. C
5. $\dfrac{4}{5}$
7. $\dfrac{5}{4}$
9. $-\dfrac{3}{2}$
11. 0
13.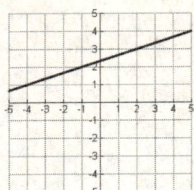
15.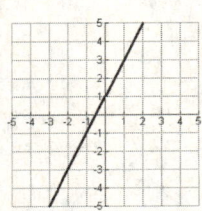
17.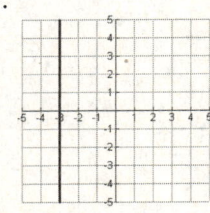
19. a) $-\dfrac{2}{7}$
    b) $\dfrac{7}{2}$

Section 7.4
1. $m = 2, (0, 3)$
3. $m = 3, (0, -2)$
5. a) $m = \dfrac{1}{5}, (0, 2)$
   b) $y = \dfrac{1}{5}x + 2$
7. a) $m = \dfrac{3}{2}, (0, 0)$
   b) $y = \dfrac{3}{2}x$
9. parallel
11. neither
13. $y = -4x + 5$
15. $y = \dfrac{1}{6}x + \dfrac{40}{3}$
17. a) $y = x + 3$
    b)

Chapter 7 Answers

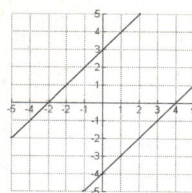

11. −25

13.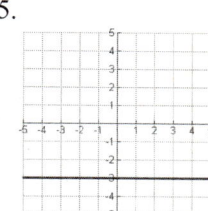

Section 7.5
1. E
3. A, B, C, D
5.
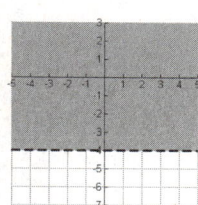

15.
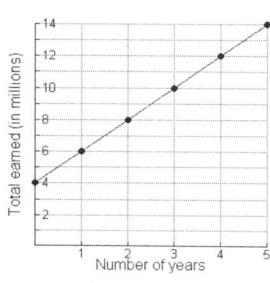

17. a) $f(x) = 4,000,000 + 2,000,000x$
b)

7.
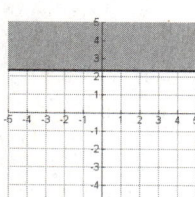

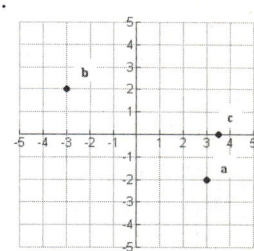

c) $12,000,000

Practice Test 7A
1. II
2. III
3. y-axis
4. I
5.

9.
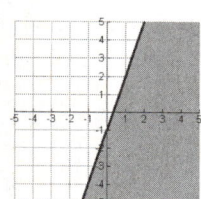

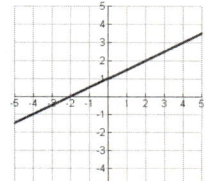

11.
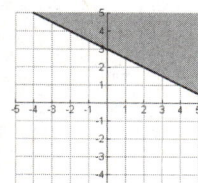

6. $(-1, -3)$, $\left(0, -\dfrac{8}{3}\right)$

7.

13. <
15. <
17. >, ≥

Section 7.6
1. a) D: $\{-8, -5, 1, 3, 54\}$
R: $\{-2, 4, 6, 12, 16\}$
b) function
3. a) D: $\{-1, 1, 7, 8\}$
R: $\{-5, -2, 0, 2, 3\}$
b) not a function
5. not a function
7. not a function
9. −11

Chapter 7 Answers

8.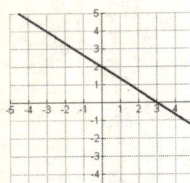

9.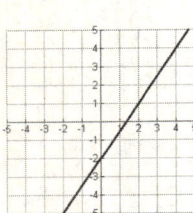

10.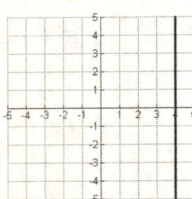

11. $x = -1$
12. $y = \dfrac{5}{2}$
13. $-\dfrac{1}{2}$
14. a) $m = 2,\ b = -2$
    b) $y = 2x - 2$
15. a) $m = -3,\ b = 0$
    b) $y = -3x$
16. $-\dfrac{3}{5}$
17. $-\dfrac{1}{6}$
18. $y = -x + 7$
19. $x = -6$
20.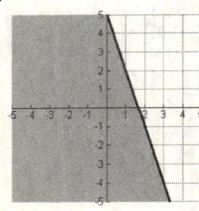

21.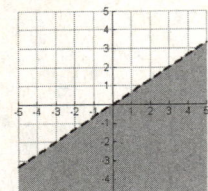

22. a) D:$\{-8, -5, 1, 3, 54\}$
       R:$\{-2, 4, 6, 12, 16\}$
    b) function
23. true
24. False, a vertical line does not represent a function.

Practice Test 7B
1. b
2. d
3. b
4. c
5. c
6. a
7. d
8. a
9. c
10. a
11. c
12. b
13. c
14. b
15. d
16. c
17. b

Chapter 8 Answers

Section 8.1
1. a
3. consistent, exactly one solution
5. dependent, infinite number of solutions
7. no solution
9. infinite number of solutions
11.

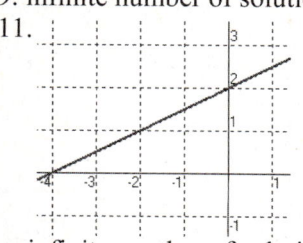

infinite number of solutions, dependent

13.

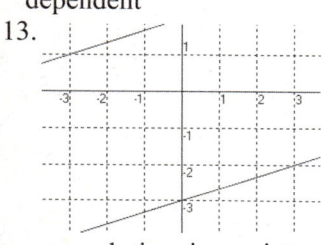

no solution, inconsistent

15.

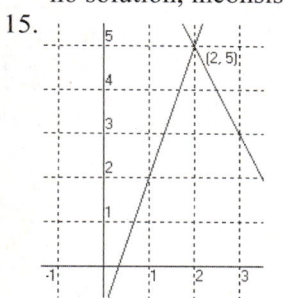

Section 8.2
1. $(2, 3)$
3. $(-5, -6)$
5. $\left(\dfrac{1}{2}, 3\right)$
7. $(-3, -4)$
9. 28, 41

Section 8.3
1. $(2, 5)$
3. $(-2, 3)$
5. $(-3, 4)$

7. $\left(\dfrac{46}{19}, -\dfrac{45}{19}\right)$
9. 4, 9

Section 8.4
1. 7.5, 4.5
3. The length is 8 ft and the width is 6 ft.
5. 455 mph
7. $3000 at 2%, $1000 at 6%
9. 3 L of 30%, 22 L of 5%

Section 8.5
1.

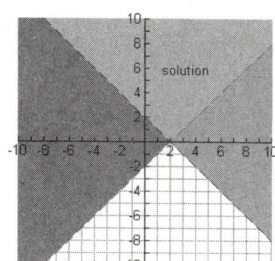

3.

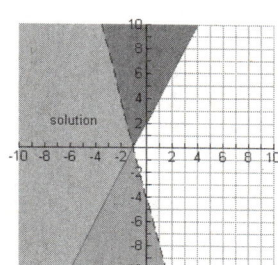

5.

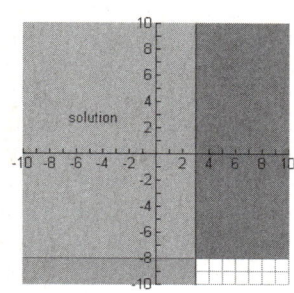

Practice Test 8A
1. $(5, -1)$
2. inconsistent, no solution
3. consistent, exactly one solution
4. no solution
5. exactly one solution
6. infinite number of solutions

7.

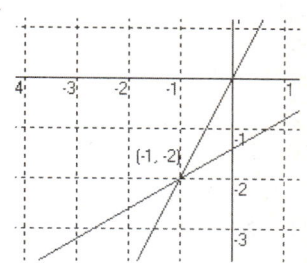

8. no solution

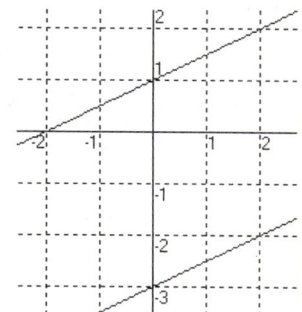

9.

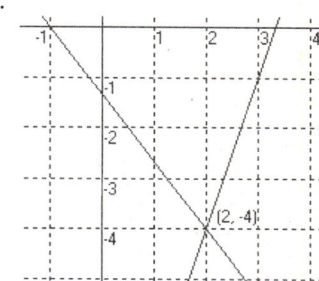

10.

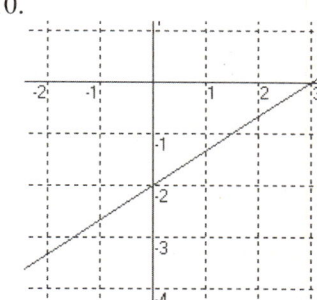

infinite number of solutions
11. $(-1, -2)$
12. $\left(\dfrac{1}{2}, -\dfrac{1}{2}\right)$
13. no solution
14. $(2, -3)$
15. infinite number of solutions

Chapter 8 Answers

16. $\left(2, \dfrac{1}{3}\right)$
17. (2, 5)
18. $(1, -1)$
19. $\left(\dfrac{3}{2}, -\dfrac{1}{2}\right)$
20. no solution
21. 425 mph
22. $3000 at 2%, $7000 at 5%
23. 16 lb of $4.30/lb, 184 lb of $1.80/lb
24.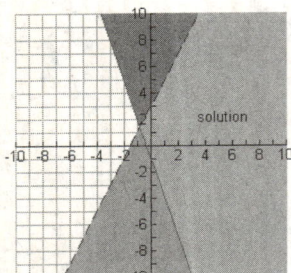
25.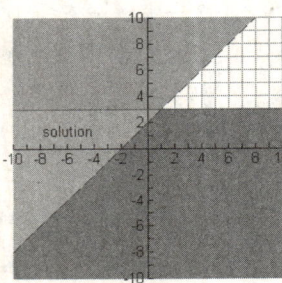

Practice Test 8B
1. d
2. a
3. b
4. a
5. b
6. a
7. a
8. c
9. b
10. d
11. b
12. d
13. a
14. c
15. a
16. c
17. b
18. d
19. d
20. c
21. a
22. b
23. c
24. c
25. a

Chapter 9 Answers

Section 9.1
1. -13
3. 11
5. -5
7. $-\dfrac{3}{8}$
9. $21^{1/2}$
11. $(18x^2)^{1/2}$
13. irrational
15. imaginary
17. true
19. false
21. false
23. $-4, \dfrac{1}{2}, 2, \sqrt{7}, 3.5, 5$

Section 9.2
1. $3\sqrt{3}$
3. $5\sqrt{7}$
5. $4\sqrt{30}$
7. $bc\sqrt{ac}$
8. $x^2yz^5\sqrt{y}$
9. $m^3n^5\sqrt{mn}$
11. $4ab^3\sqrt{14}$
13. 13
15. 18
17. $3x^3y^3\sqrt{15}$
19. $3x$
21. $5xy\sqrt{2x}$

Section 9.3
1. $4\sqrt{6}$
3. $3\sqrt{x}+5$
5. $5\sqrt{6}$
7. $-5\sqrt{2}$
9. $5\sqrt{a}-5\sqrt{10}$
11. $27-\sqrt{3}$
13. 31
15. area = $9-4\sqrt{5}$; perimeter = $4\sqrt{5}-8$

Section 9.4
1. 4
3. $\dfrac{5}{13}$
5. $\dfrac{b\sqrt{abz}}{2z}$
7. $\dfrac{\sqrt{15}}{5}$
9. $\dfrac{\sqrt{xy}}{y}$
11. $\dfrac{4\sqrt{5x}}{xy^2}$
13. $\sqrt{7}+\sqrt{2}$
15. $\dfrac{27\sqrt{x}+9x}{9-x}$

Section 9.5
1. 16
3. 169
5. no solution
7. 7
9. $4, 16$
11. 5
13. $x=5$

Section 9.6
1. 10
3. $4\sqrt{6} \approx 9.80$
5. 5
7. $0.5\pi \approx 1.57$ sec

Section 9.7
1. -4
3. 4
5. $3\sqrt[4]{5}$
7. $x^{5/3}$
9. y
11. 2
13. $\sqrt[3]{x}$
15. $\sqrt[5]{a}$
17. 216
19. $x^{8/15}$

Practice Test 9A
1. $(7y)^{1/2}$
2. $\sqrt[5]{y^3}$
3. 8
4. $2\sqrt{15}$
5. $3a\sqrt{6}$
6. $4x^2y^2\sqrt{6x}$
7. $9xy^2\sqrt{5y}$
8. $3xy^3\sqrt{2x}$
9. $\dfrac{1}{6}$
10. $\dfrac{36-12\sqrt{x}}{9-x}$
11. $\dfrac{7\sqrt{11}}{11}$
12. $\dfrac{5\sqrt{7x}}{7}$
13. $\dfrac{3\sqrt{3x}}{xy^2}$
14. $\dfrac{36-12\sqrt{x}}{9-x}$
15. $5\sqrt{5}+10$
16. $10\sqrt{6}$
17. 0
18. 5
19. $-1, 3$
20. $5\sqrt{3} \approx 8.66$
21. $4\sqrt{5} \approx 8.94$
22. $\dfrac{1}{64}$
23. x^2
24. 75.05 ft per sec
25. 8 ft

Chapter 9 Answers

Practice Test 9B
1. b
2. c
3. d
4. a
5. a
6. c
7. b
8. b
9. d
10. c
11. c
12. c
13. a
14. b
15. c
16. a
17. d
18. d
19. c
20. c
21. a
22. b
23. b
24. a
25. c

Chapter 10 Answers

Section 10.1
1. ± 15
3. $\pm 2\sqrt{2}$
5. $-6, 12$
7. $\dfrac{-6 \pm 3\sqrt{5}}{7}$
9. length ≈ 49.30 ft, width ≈ 30.43 ft

Section 10.2
1. $-8, 1$
3. $4, 6$
5. $\dfrac{-7 \pm \sqrt{77}}{2}$
7. $\dfrac{-1 \pm \sqrt{15}}{7}$
9. $-8, -5$

Section 10.3
1. two distinct real number solutions
3. one real number solution
5. two distinct real number solutions
7. $-5, 8$
9. $\dfrac{7 \pm \sqrt{109}}{6}$
11. no real number solution
13. $\dfrac{-3 \pm \sqrt{29}}{2}$
15. 2.36 sec, 5.64 sec

Section 10.4
1. $x = 0$, $(0, 4)$, downward
3. $x = -1$, $(-1, 0)$, downward
5.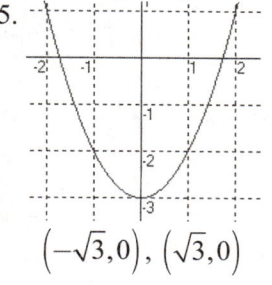
$(-\sqrt{3}, 0), (\sqrt{3}, 0)$

7.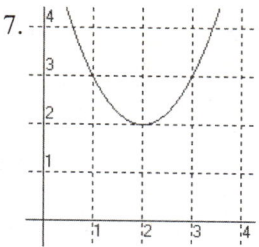
no x-intercepts

9.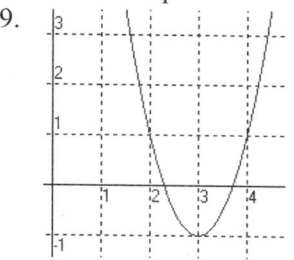
$\left(\dfrac{6-\sqrt{2}}{2}, 0\right), \left(\dfrac{6+\sqrt{2}}{2}, 0\right)$

11.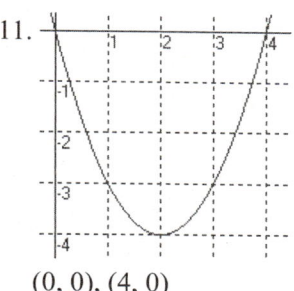
$(0, 0), (4, 0)$

13.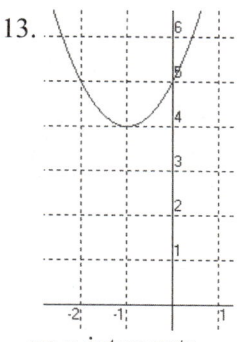
no x-intercepts

Section 10.5
1. $7i$
3. $3i\sqrt{6}$
5. $-3 - i\sqrt{5}$
7. $11 + 7i$
9. $4 + 5i$
11. $-8i$

13. $\pm 4i\sqrt{2}$
15. $2 \pm 3i$
17. $\dfrac{5 \pm 2i\sqrt{2}}{6}$

Practice Test 10A
1. $\pm 2\sqrt{5}$
2. $\dfrac{-5 \pm \sqrt{10}}{4}$
3. $4 \pm 3\sqrt{2}$
4. $2, 3$
5. $\dfrac{1}{2}, \dfrac{3}{4}$
6. $\dfrac{-7 \pm \sqrt{97}}{6}$
7. $5 \pm 5\sqrt{3}$
8. Answers will vary.
9. two distinct real solutions
10. no real number solution
11. $x = 1$
12. $x = 3$
13. upward
14. downward
15. $(1, 4)$
16. $(2, 2)$
17.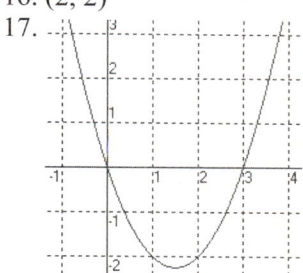
$(0, 0), (3, 0)$

18.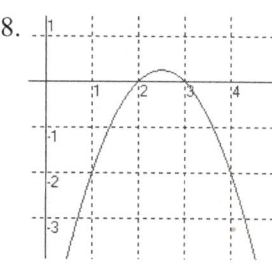
$(2, 0), (3, 0)$

Chapter 10 Answers

19.

$\left(2-\sqrt{3}, 0\right), \left(2+\sqrt{3}, 0\right)$

20. 13, 15
21. 3.75 sec
22. $-2 - i$
23. -1
24. $-5 \pm 7i$
25. $\dfrac{5 \pm 3i\sqrt{6}}{2}$

Practice Test 10B
1. a
2. b
3. b
4. d
5. c
6. c
7. a
8. c
9. b
10. a
11. a
12. c
13. a
14. b
15. a
16. b
17. d
18. b
19. c
20. c
21. a
22. c
23. b
24. d
25. a